T0255389

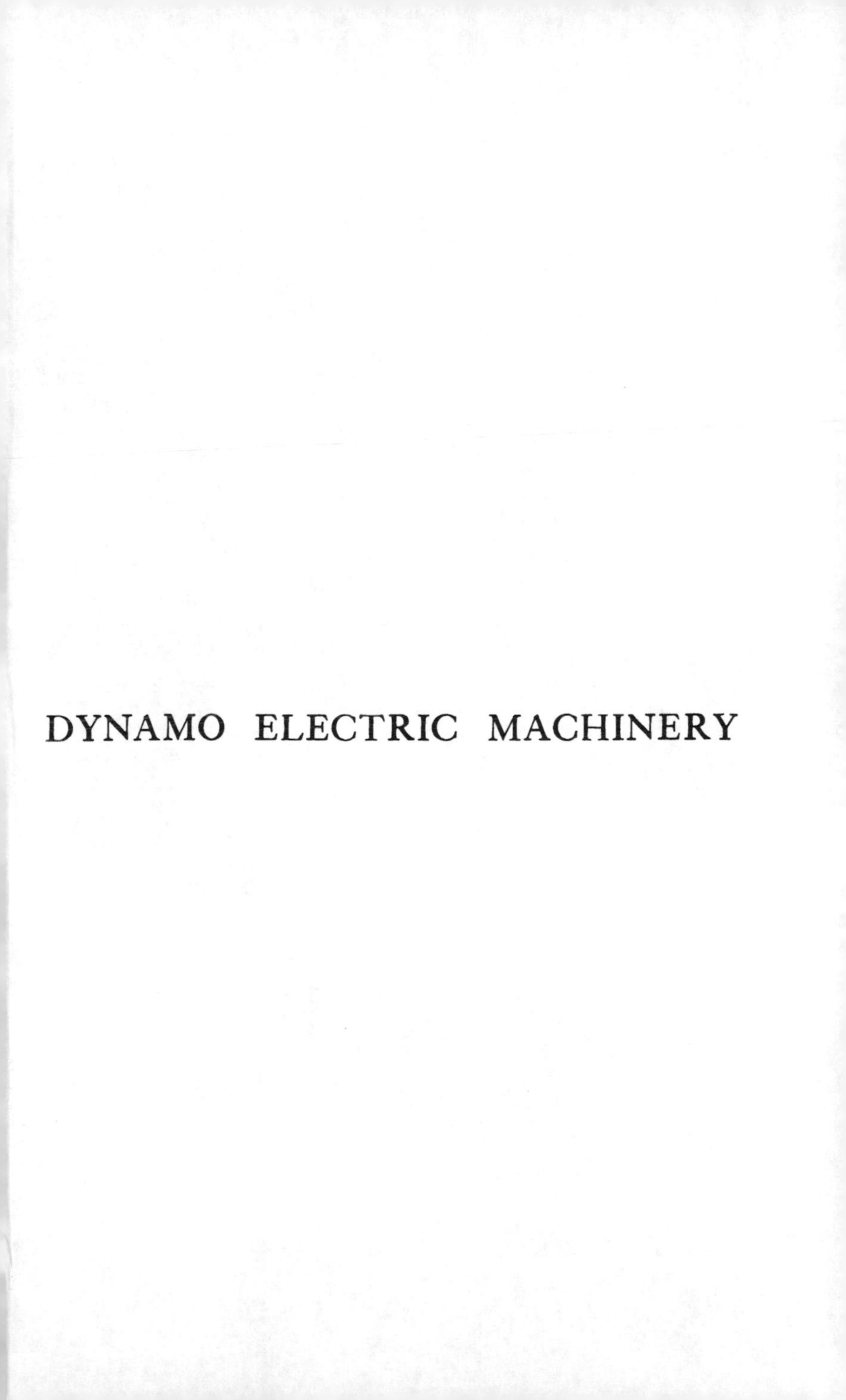

DYNAMO ELECTRIC MACHINERY

THE ELEMENTARY THEORY
OF
DIRECT CURRENT
DYNAMO ELECTRIC MACHINERY

by

C. E. ASHFORD, M.A.

Head Master, Royal Naval College, Dartmouth

and

E. W. E. KEMPSON, B.A.

Assistant Master, Royal Naval College, Dartmouth

Cambridge
at the University Press
1908

CAMBRIDGE
UNIVERSITY PRESS

University Printing House, Cambridge CB2 8BS, United Kingdom

Cambridge University Press is part of the University of Cambridge.

It furthers the University's mission by disseminating knowledge in the pursuit of education, learning and research at the highest international levels of excellence.

www.cambridge.org
Information on this title: www.cambridge.org/9781107494930

© Cambridge University Press 1908

This publication is in copyright. Subject to statutory exception and to the provisions of relevant collective licensing agreements, no reproduction of any part may take place without the written permission of Cambridge University Press.

First published 1908
First paperback edition 2015

A catalogue record for this publication is available from the British Library

ISBN 978-1-107-49493-0 Paperback

Cambridge University Press has no responsibility for the persistence or accuracy of URLs for external or third-party internet websites referred to in this publication, and does not guarantee that any content on such websites is, or will remain, accurate or appropriate.

PREFACE

THE authors of this little book believe that in the training of an Electrical Engineer there should be included a knowledge of the theory of the subject built up logically from first principles, each step being illustrated with the help of some piece of machinery or practical appliance of a general and simple rather than an elaborate or necessarily up-to-date type; and that he should be taught not to accept statements without evidence of their truth. If he follows this plan he will be able to understand any piece of electrical machinery which he comes across, not because he has ever seen or heard of it before, but because he can recognise it is a particular case of a general class whose fundamental principles he has mastered. If on the other hand he attempts to learn one by one all the various commercial machines and appliances on the market at the moment, his task will be unending. The creation of new types is unceasing, but the principles are permanent.

Of the elementary text-books dealing with continuous current dynamo electric machines, the majority dismiss the theory of the subject with a few brief statements and worked examples, and devote many chapters to the discussion of special forms of machines together with full explanations of their mechanical details; or if they do attempt a more extended investigation of the theory, it is of a disjointed nature and contains many statements of propositions which it is said may easily be proved, but whose proofs are omitted. The latter is a procedure essentially bad from an educational standpoint; the former would be justifiable were the intention to provide a handbook to workshop practice for students familiar with the theory.

The present book does not cover much ground, but the authors have attempted to present as logical a treatment as is possible in so elementary a work. It is of course intended to be used only as a note book accompanying a course of experimental lectures; many paragraphs require amplification, this being especially the case where the treatment is of the generally accepted type. Where however the treatment differs from that usually adopted, as in the chapter on the magnetic circuit, the notes have been given more fully.

If it is desired to present a still easier course to the student, omitting many proofs and giving merely a descriptive treatment of the subjects, the following paragraphs may be read:—

Articles 1—34, 39—44, 56—58; or for a still shorter course, Articles 1—19, 27—33, 56—58.

C. E. A.
E. W. E. K.

September, 1908.

CONTENTS

CHAPTER I

RELATIONS BETWEEN ELECTRIC CURRENTS AND MAGNETIC FIELDS

1. Magnetic Effect of a Current of Electricity.

If we pass a current of say 40 ampères down a vertical wire, which pierces a horizontal strip of cardboard, and sift iron filings on the cardboard, the filings arrange themselves in concentric circles and those filings which are nearer the wire appear to be directed with greater force.

If we cause the current to flow *downwards* in the conductor we find that a small exploring magnet placed on the cardboard will set itself along a tangent to the circle whose centre is in the wire conductor, and that viewed from above the 'north' pole of the magnet points in a *clockwise* direction round the conductor.

If the current in the conductor is reversed the 'north' pole of the magnet points in the opposite direction.

Now by convention the exploring magnet referred to sets itself so that its 'north' pole points in the direction of the lines of force. It follows therefore that the direction of the lines of force round a current-bearing conductor is such that if we look along the conductor in the direction in which the current is flowing the lines of force appear clockwise in direction.

The field of force which is set up by a current-bearing conductor which is bent into the form of a ring can be examined in a similar way by using iron filings and an exploring magnet.

The figure shows a section of such a field. Each line of force passes once through the ring and is completed outside.

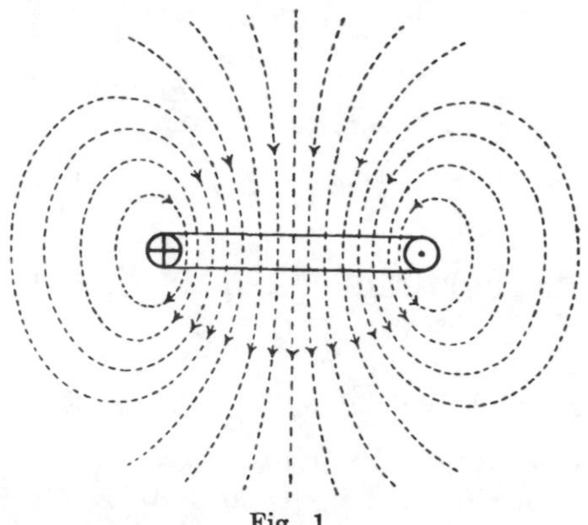

Fig. 1.

[In this and subsequent figures a conductor carrying a current downwards is represented in section by a circle with a cross in it and a conductor carrying a current upwards by a circle with a dot in it. The cross is supposed to represent the feathers of an arrow going away from the observer and the dot to represent the point of the arrow coming towards the observer.]

The field is like that due to a short disc-shaped magnet of large cross section with its edge on the conductor, the 'north' pole extending all over one face, the 'south' pole all over the other.

On looking at the ring, if the current appears to travel clockwise the lines of force appear to go away from the eye, and if anti-clockwise towards the eye.

The magnetic effect is increased by joining a number of coils in series in the form of a 'solenoid,' and the next figure represents the section of the field set up by such a solenoid carrying a current.

The field is similar to that due to an ordinary bar magnet.

If a soft iron bar is inserted in the solenoid with its axis coincident with that of the solenoid the magnetic effect of the solenoid is vastly increased, as though the current found it easier to set up a magnetic field in iron than in air. The iron becomes a magnet with its 'north' pole towards B.

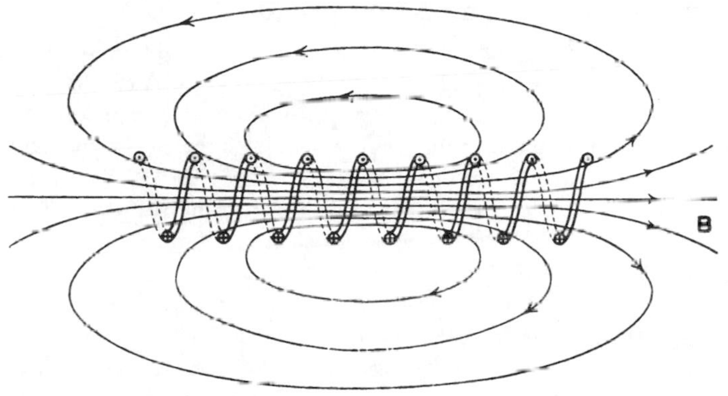

Fig. 2.

A magnet of this kind is called an electro-magnet. If the current is stopped the magnetism in the soft iron almost disappears, and if the current is reversed the magnetism in the iron is also reversed.

A convenient trick for finding the polarity of an electro-magnet is given by the following figures:

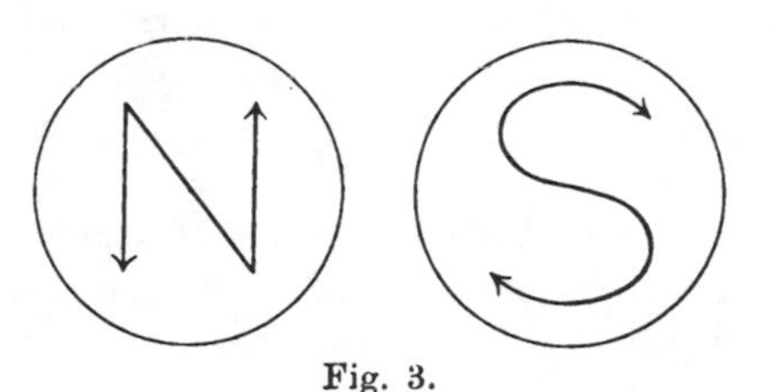

Fig. 3.

The arrows show the direction of the current when looking at the end of the solenoid and the letter gives the polarity of that end of the magnet.

The 'Corkscrew Rule' is another trick for the same purpose. The direction of rotation of the corkscrew gives the direction of the current, and the forward motion of the corkscrew that of the induced lines of force.

2. Magnetisation of Iron and Steel.

If we pass a current through a solenoid of wire and insert, first a soft iron core, and then a hard steel core of the same dimensions, we find that the magnetism induced in the soft iron is very much stronger than that which is induced in the hard steel by the same current, but that on breaking the electric circuit the whole or nearly the whole of the magnetism disappears from the soft iron, while a large proportion of the induced magnetism remains in the hard steel.

If we pass a current down a long thin solenoid of wire we can measure the strength of the magnetic field produced by means of a magnetometer placed near the end of the solenoid, using the same method as that employed to measure the strength of a long permanent magnet.

Now if we plot a curve of which the ordinates represent a measure of the strength of the magnetic field produced by the solenoid and the abscissae the strength of the current in the solenoid, and repeat the process for the same solenoid, but with a soft iron core and a steel core, then we shall be able to compare the magnetic effect produced in air, soft iron and steel for the same expenditure of electrical power and otherwise under the same conditions.

A simple and very rough experiment may be carried out as follows:

Take three exactly similar solenoids of insulated wire, one of which is wound on a core of wood or glass, either of which is magnetically equivalent to air, the other two being wound on cores of soft iron and steel.

If there is any residual magnetism in the iron or steel it must be destroyed.

First connect the solenoid, with the wooden core, in series

with a commutator, battery of cells and variable resistance, and arrange a magnetometer as shown in the figure.

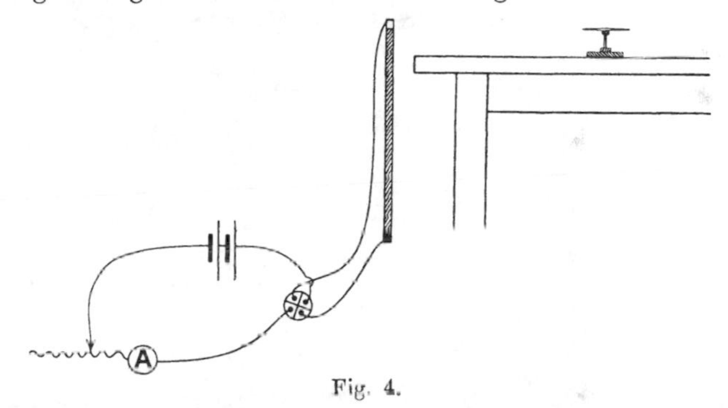

Fig. 4.

Increase the current in the solenoid gradually and note down from time to time the current and the tangent of the angle of deflection of the magnetometer.

It will be found that the tangent of the angle of deflection of the magnetometer, and therefore the strength of the magnetic field produced by the solenoid, is directly proportional to the current, also that the magnetic effect is the same whether the current is being increased or diminished, and when the current is stopped the whole of the magnetic effect disappears.

The curve connecting magnetic field strength with current strength is in fact a straight line through the origin.

Now substitute the coil with the soft iron core for that with the wooden core and see that it occupies exactly the same position.

Gradually increase the current from zero and again note down the values of the current and the tangent of the angle of deflection of the magnetometer.

The general shape of the curve connecting these two quantities will be found to resemble closely that of the curve in the diagram on the next page.

The curve may be roughly divided into three parts OA, AB and CD.

Between O and A the magnetic effect increases slowly.

Between A and B the magnetic effect is practically proportional to the current and increases rapidly with the current.

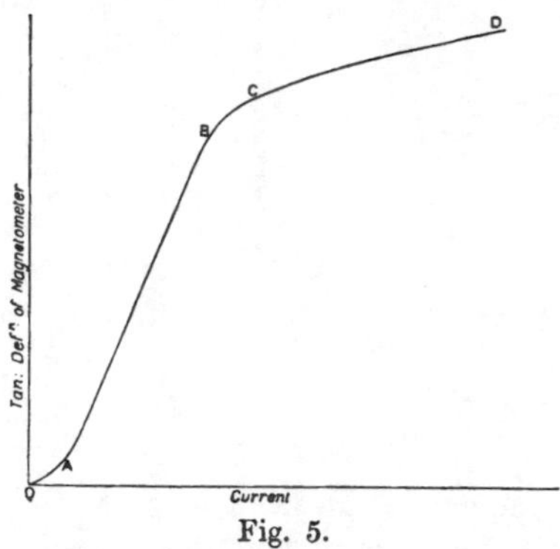

Fig. 5.

Between C and D the magnetic effect increases more or less uniformly but very much more slowly than before. There is, of course, a transition stage represented by the part of the curve BC where the magnetic effect does not increase at all uniformly.

When the iron is in a state corresponding to a point on the curve between C and D it is said to be saturated.

The part of the curve BC is sometimes referred to as the 'knee' of the magnetisation curve.

Now diminish the current gradually to zero, commutate, so that the current will flow the opposite way round the solenoid, again increase the current until the iron becomes nearly saturated with reverse magnetism, diminish the current to zero, again commutate and again increase the current.

The complete curve connecting the two quantities is represented in Fig. 6. The full line shows the curve obtained for soft iron and the dotted line that for hard steel.

When once the iron or steel has been magnetised, if the strength of the current in the solenoid is altered, the metal has a greater or less tendency to remain in the same magnetic condition in which it was before the alteration in the current; for example, when the current in the solenoid is reduced to zero the magnetism does not all disappear, and this is more noticeable in the case of steel than of iron*. This lagging of the effect behind the cause which produces it is called *hysteresis*.

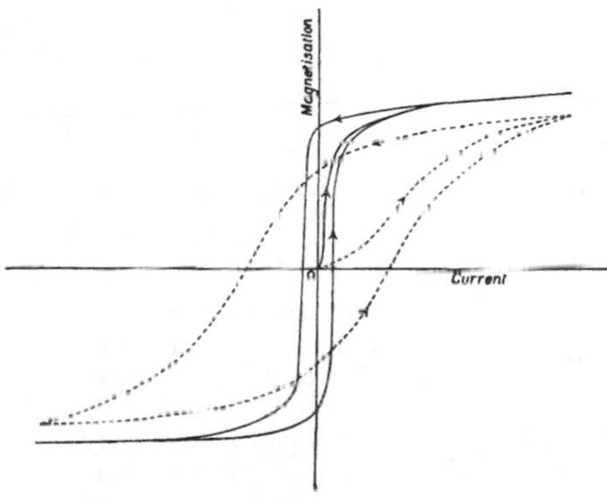

Fig. 6.

It can be shown that the work done in magnetising and demagnetising a piece of iron or steel through a complete cycle is proportional to the area of the closed magnetisation curve, or hysteresis curve as it is sometimes called.

This is of great importance in the construction of Dynamos and Motors, as will be seen later.

It will be proved later that the magnetising effect of a coil carrying a current is proportional to the product of the current and number of turns in the coil, or to the 'ampère turns' as this product is called.

To verify this statement roughly, magnetise an iron rod by means of a coil of 'S' turns carrying a current of 'C' ampères.

* This is apparently contradicted by Fig. 6; but these curves represent the results of measurements on long bars of iron and steel protected from the least vibration. If the bars had been short, or had been tapped when the magnetising current was reduced to zero, practically the whole of the magnetism would have disappeared from the soft iron, while the steel would have retained its magnetism.

It will be found that if we change the coil and use $\frac{S}{N}$ turns, a current of CN ampères is required to produce the same effect.

The iron rod must of course be completely demagnetised before each part of the above test.

3. Force on a Current-bearing Conductor in a Magnetic Field.

If a current-bearing conductor, which is free to move, is placed in a magnetic field so that it lies at right angles to the direction of the lines of force in the magnetic field, it will move parallel to itself and so as to cut the lines of force at right angles.

This can be illustrated by suspending a wire between the poles of a horse-shoe magnet so that the lower end of the wire dips into a pool of mercury. On passing a current down the wire, the wire will be driven out from between the poles. If the current is reversed the wire will be driven in the opposite direction.

To understand this we will consider the effect on a previously straight parallel magnetic field of inserting a current-bearing conductor in it.

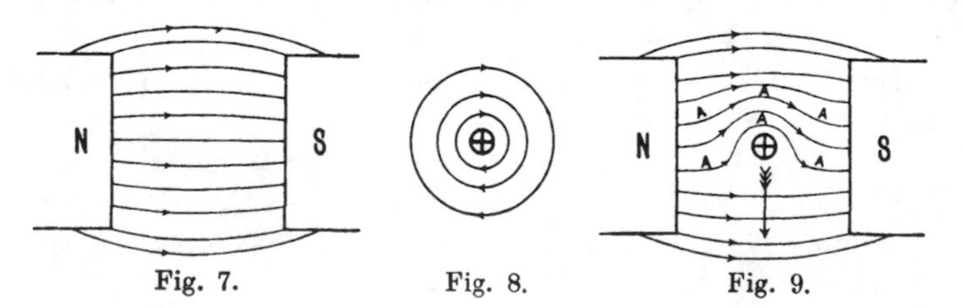

Fig. 7. Fig. 8. Fig. 9.

Fig. 7 represents a field of force consisting of parallel or nearly parallel straight lines such as is obtained between the poles of a large magnet.

Fig. 8 represents the field which is set up by a current-bearing conductor.

Fig. 9 represents the field which is obtained by superimposing the field of Fig. 8 on the field of Fig. 7.

The nature of the magnetic field in each of these three cases can be easily demonstrated in the usual way with iron filings and a small exploring magnet.

It is a matter of experience that whenever a line of force joins two bodies, those two bodies are subjected to a force in the shape of a pull which tends to bring them together along the line of force, just as though the line of force were an extended elastic string, with this difference that as the bodies come nearer together the force becomes greater instead of less, as would be the case if the bodies were joined by elastic strings.

If we further suppose that lines of force act like elastic strings, even on bodies by which they are deflected, then reference to Fig. 9 will make it clear that lines of force such as A, A, A being compressed together tend to push the conductor away down the paper in the direction indicated by the large arrow.

Although the above argument is not of the nature of strict reasoning, still its application is always easy and happens to lead to correct results. Two illustrative examples are given below.

1. Two parallel conductors carrying current in the same direction.

The magnetic field set up by such an arrangement is like that represented in section in the accompanying diagram, Fig. 10.

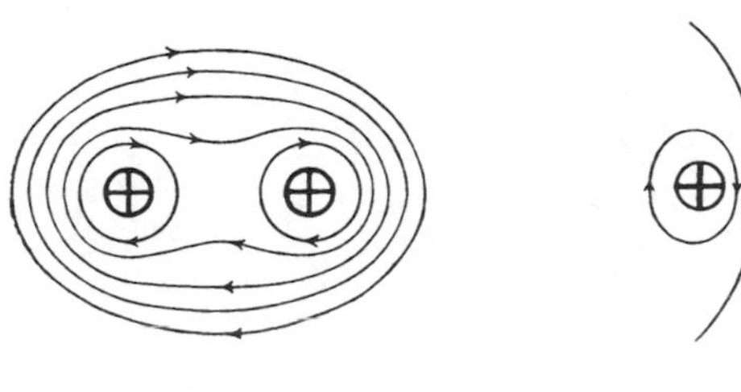

Fig. 10.

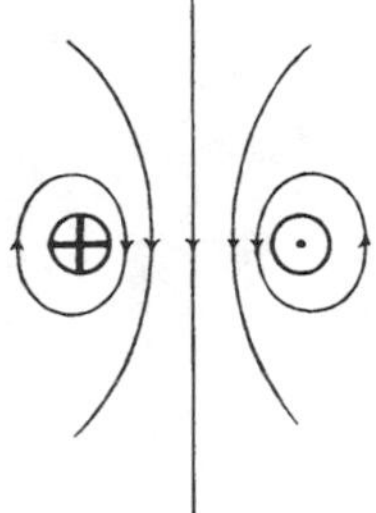

Fig. 11.

From the nature of the field we expect the conductors to be forced towards one another and a simple experiment shows that they are.

2. Two parallel conductors carrying current in opposite directions.

A diagram of the field is given in Fig. 11 above. The conductors are forced away from one another.

This force which acts upon a current-bearing conductor placed in a magnetic field is made use of in motors for the conversion of electrical energy into mechanical energy.

CHAPTER II

ESSENTIAL PARTS OF DYNAMO ELECTRIC MACHINES

4. Suspended coil Galvanometer.

If a coil carrying a current is suspended in a magnetic field in such a way that the plane of the coil is parallel to the direction of the lines of force in the magnetic field, a couple comes into action which tends to turn the coil round so that its plane is perpendicular to the direction of the lines of force.

The suspended coil galvanometer consists of a coil of many turns of fine insulated wire suspended in the manner described between the poles of a strong permanent magnet.

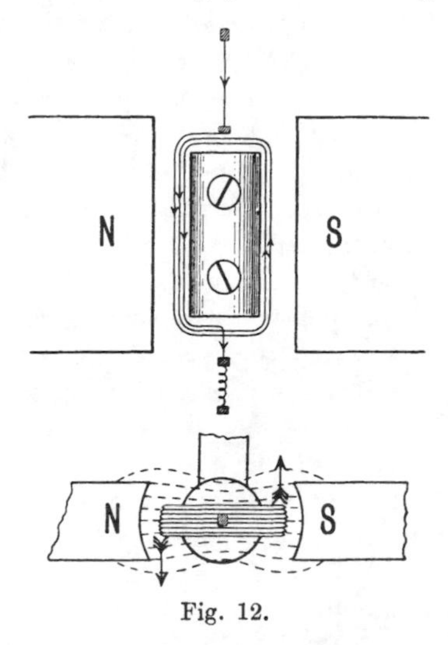

Fig. 12.

When a current is passed through the coil a couple comes into action which turns the coil round, and in so doing has to overcome the torque due to twist in the fine wire by which the coil is suspended. The resisting torque of the suspending wire is proportional to the angle of twist.

In the position of equilibrium the resisting torque of the suspending wire is equal to the turning couple of the coil, and this for small angles of twist is proportional to the current.

Hence the angle of twist of the coil is proportional to the current.

A soft iron core is introduced inside the coil in order to strengthen the field in which the coil lies; it also has the effect of making the lines of magnetic force in the gap lie parallel to the plane of the coil throughout a large angle of twist.

5. Motors. First Notion.

In its simplest form a motor consists of a single coil of wire arranged as in the galvanometer, but now the coil is free to rotate.

Each end of the coil is attached to a hemicylindrical strip of copper called a commutator strip.

Fig. 13 shows the arrangement of the coil and magnets.

An electric current is led into the coil by means of the brushes, connections being made as indicated in the diagram.

In the position shown, current is led into the coil through the brush E and commutator strip H round the coil in the direction DCBA and so back to the battery. As in the galvanometer the coil is acted on by a torque—we will now consider this in detail.

The conductors AB and CD are acted on by forces in opposite directions at right angles to the lines of force, as may be seen by reference to the diagram in elevation above. If the magnetic field is uniform these forces will be the same in whatever position the coil may be, but the two forces form a couple which varies from a maximum, when the plane of the coil is parallel to the direction of the lines of force, to zero, when the plane of the coil is perpendicular to the direction of the lines of force.

The coil will therefore rotate in a clockwise direction until it comes into the plane *pp* and will remain there unless the direction of the current or of the magnetic field is reversed.

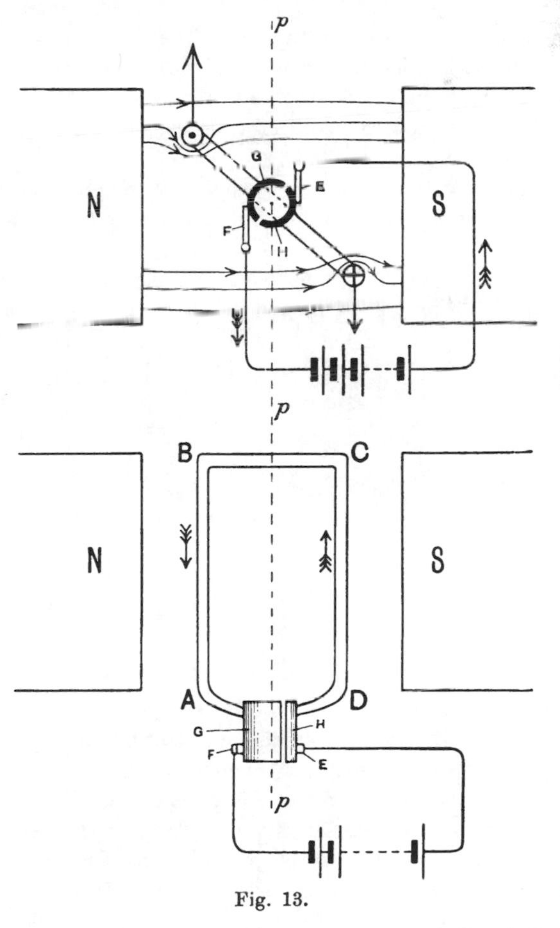

Fig. 13.

The coil is, however, carried a little beyond this position by its own momentum and then the commutator strip G leaves the brush F and comes in contact with the brush E while the commutator strip H comes in contact with the brush F.

The current in the coil is thus reversed and the forces acting on the conductors AB and CD are also reversed, thus again producing a clockwise couple on the coil. The coil then continues to rotate in the same direction since the couple always acts in the same direction though it fluctuates in magnitude.

The behaviour of the coil may be considered from another point of view.

We have seen that as far as its effect on neighbouring bodies is concerned a coil carrying a current is equivalent to a short disc-shaped magnet.

The coil of the simple form of motor which we have considered may therefore be looked upon as a magnet and will

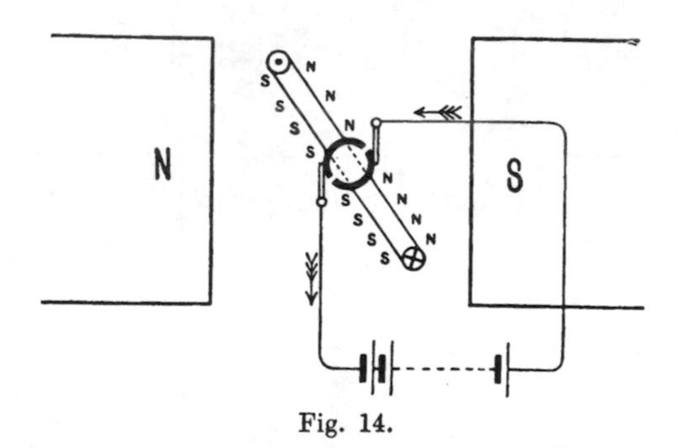

Fig. 14.

simply rotate until its 'north' face comes opposite to the 'south' pole of the permanent magnet and its 'south' face opposite the 'north' pole of the permanent magnet.

As soon as the coil arrives in this position, the direction of the current in the coil is reversed by the action of the commutator and this reverses the polarity of the coil considered as a magnet so that the motion becomes continuous.

6. Armatures.

A coil such as we have described may be used to produce

continuous rotation, but the couple generated is weak and fluctuates from zero to a maximum twice in every revolution.

In a practical form of motor a large number of such coils are used in series with one another and spaced evenly round a soft iron core.

The soft iron core serves as a drum on which to wind the coils and also by diminishing the air space between the poles of the field magnets facilitates the generation of a powerful field across the conductors.

By using a large number of coils the driving couple is correspondingly increased, and by spacing the coils evenly round the core a uniform couple can be obtained, there being always several of the coils with their planes parallel to the direction of the magnetic lines of force.

There will of course be a number of commutator strips corresponding to the number of coils, and each commutator strip may form a junction between two adjacent coils.

As each coil comes into the plane perpendicular to the lines of force it must of course be 'commutated' and the brushes are so placed as to bring about this result.

Any such coil or system of coils is called an 'armature,' due probably to the fact that it lies between the poles of the field magnets like the 'armature' or 'keeper' of a permanent horseshoe magnet.

7. Field Magnets.

In the majority of cases the field magnets are electro-magnets.

In a 'shunt' motor the current round the exciting coils of the magnets is taken from the same source of E.M.F. as the current through the armature, but the coils themselves are in parallel with or form a shunt to the armature circuit.

In a 'series' motor the same current flows through the armature and field magnet circuit, the two circuits being in series.

A diagram is given which shows the connections for the two types of winding.

In the 'shunt' type of winding a large number of turns of fine wire are used, the circuit has a high resistance and a small current is taken.

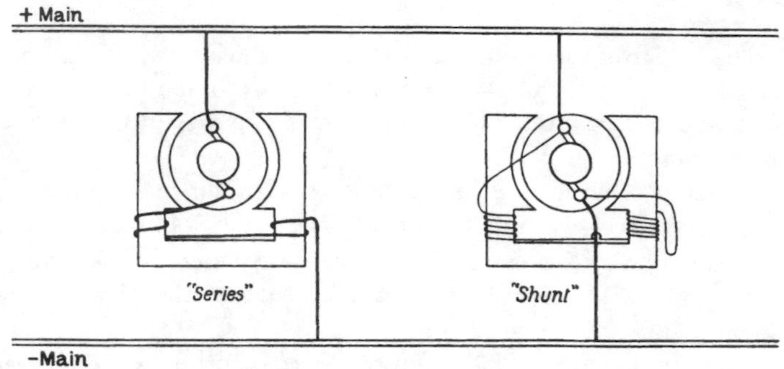

Fig. 15.

In the series type of winding the whole current must flow round the field circuit, which therefore consists of a comparatively few turns of thick wire having a low resistance.

As far as the strength of the magnets is concerned it does not matter whether the exciting circuit consists of a large number of turns of wire carrying a small current, or of a few turns of wire carrying a large current, provided the product of the current and number of turns is the same in both cases.

The pole pieces of the field magnets are shaped so as to lie very close to the armature, allowing only just sufficient clearance for the armature to rotate without touching the poles of the field magnets.

8. Induction of Electric Currents.

We can show that if a conductor is caused to move in a magnetic field so as to cut lines of force, then a current is induced in the conductor provided the conductor forms part of a closed circuit. As a matter of fact whenever a conductor moves across lines of force an electromotive force is induced in the conductor whether the conductor forms part of a closed circuit or not;

the magnitude of the current (by Ohm's law) varies inversely as the resistance of the complete circuit as well as directly as this electromotive force.

To illustrate this connect up a wire to a galvanometer, and pass the wire quickly between the poles of a strong horse-shoe magnet—a current is detected.

Pass the wire back again and a current is again detected, but in the opposite direction.

Now bring the wire towards some other conductor which is carrying a steady current. A current is detected by the galvanometer and on separating the wires again a reverse current is detected.

The effect is increased in each case by increasing the strength of the magnet or by using one side of a coil instead of a single wire.

Again, if we place two conductors side by side and pass a current down one of them a momentary current can be detected in the other, but in the opposite direction. On stopping the current in the first conductor a momentary current can again be detected in the other, but this time in the forward direction.

9. Direction of Induced Current.

The above are experimental facts and do not admit of an explanation, but in order to arrive at a rule for discovering the direction of the induced current let us again assume that the lines of force act like elastic strings, which bulge out where a conductor pushes them (this bulge being in fact produced by the effect on the original field of the superimposed field of the current induced in the conductor). Then the bulge gives the direction of the superimposed field and (by the corkscrew rule) of the induced current.

For example, suppose Fig. 16 shows a conductor moving across a field and bulging the lines of force, this distorted field is produced by the combination of the fields represented in Figs. 17 and 18.

The field in Fig. 18 is due to a current *down* the conductor, and so a 'down' current is induced in the conductor in Fig. 16.

We will assume that lines of magnetic force behave in this way when cut by a conductor capable of carrying a current, but not at first doing so.

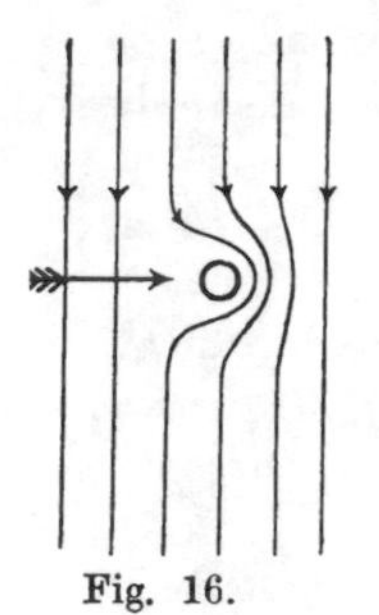

Fig. 16.

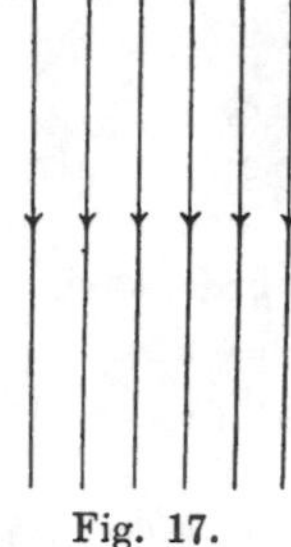

Fig. 17.

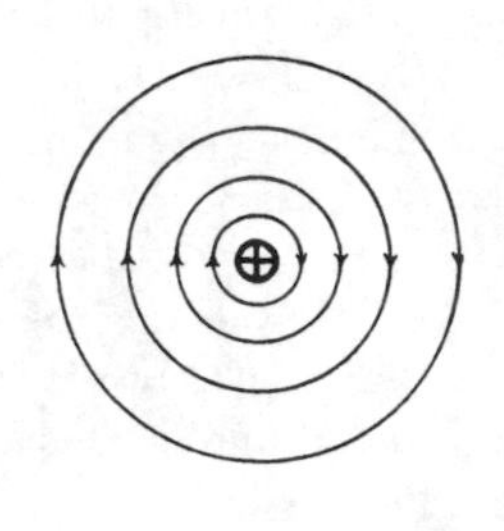

Fig. 18.

This rule may be applied to any case where a conductor moves relatively to a magnetic field of force.

Take the case of two conductors A and B lying side by side. If a current is started in A a circular field of force is generated from A and travels outwards as the current increases from zero to its full value. Ultimately a large number of lines of force will lie outside B, but to get there they must cut B, and in doing so they set up an induced current in B in the opposite direction to that in A.

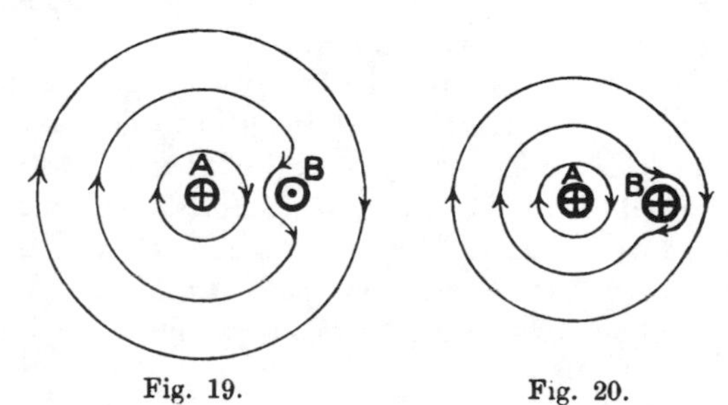

Fig. 19. Fig. 20.

Fig. 19 shows the momentary effect of starting a current in A; Fig. 20 the momentary effect of stopping the current in A.

The application of the rule to the case of a conductor which is moved parallel to the lines of force in a magnetic field, shows that there is no induced current in this case, and an easy experimental test verifies this conclusion.

Hitherto we have spoken of the induction of current by the passage of a conductor across a magnetic field, but properly speaking the effect of this motion of the conductor across a magnetic field is the induction of an electromotive force, and this electromotive force is induced whether there is a complete circuit or not, and the current follows if the circuit is complete. Again, the direction of this induced electromotive force may be such as to oppose the flow of an independent current in the conductor, and in this case the result will be a diminution of the current, but not necessarily a reversal of the current.

The foregoing paragraph is very important and its meaning should be thoroughly grasped before proceeding further.

10. Dynamo. First Notion.

A machine which generates an electromotive force by the motion of one or more conductors across a magnetic field is called a Dynamo. In its simplest form a Dynamo consists of a single coil of wire attached to a commutator and arranged so as to rotate between the poles of a magnet.

The diagrams show the arrangement of the coil and magnets. These should be compared with the diagrams on page 13.

In the position shown in Figure 21 the conductor AB is cutting lines of force in such a way that an E.M.F. is being induced which tends to send a current from A to B. The conductor CD is cutting lines of force in such a way that an electromotive force is being induced which tends to send a current from C to D. The conductors AB and CD are joined in series by means of an end connection BC and by an external circuit which is joined to the commutator strips G and H by brushes E and F.

In this position then a current flows from the brush E round the external circuit to the brush F and round the coil in the direction ABCD.

When the conductors AB—CD reach the vertical plane PP

they will both be moving parallel to the lines of force and so will not have any E.M.F. induced in them, and at this moment the electromotive force and current in the circuit will both be zero.

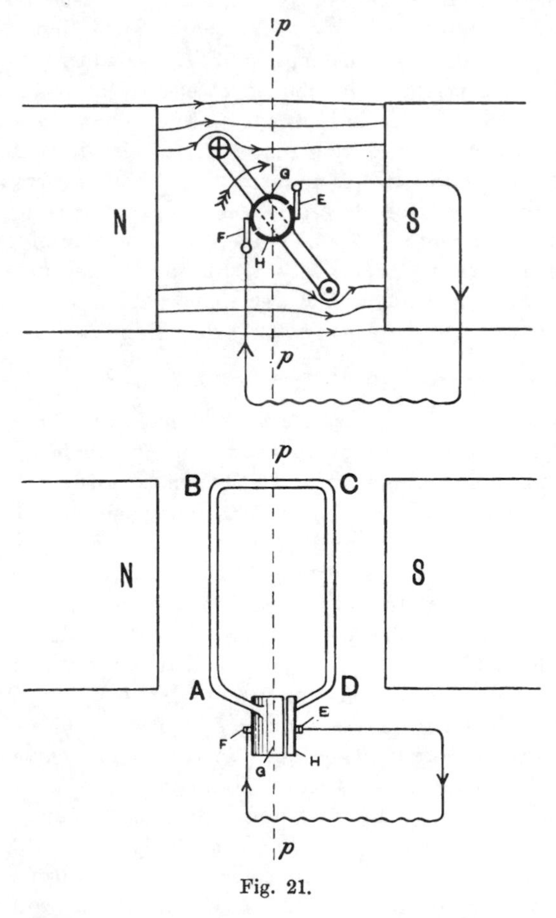

Fig. 21.

As soon as the coil passes this position the conductors AB and CD both begin to cut lines of force in the opposite direction, and hence the induced E.M.F. tends to send a current in the direction DCBA.

However, by this time the commutator strip H has left the brush E and come into contact with the brush F, and the commutator strip G has come into contact with the brush E, so that although the E.M.F. is reversed in the coil, the brush E is still touching the positive commutator strip, and the current in the external circuit does not change in direction, though it fluctuates in magnitude between zero and a maximum value twice in every revolution.

Every such coil rotated in the manner described is like a battery of two voltaic cells in series, in so far as it is a seat of electromotive force and has an internal resistance, and any number of such coils may be connected in series or parallel.

Just as cells are connected in series if a high E.M.F. is required and in parallel if a low internal resistance is required, so with the coils of a dynamo.

The practical form of the dynamo is like that of the motor, and as will be seen later the same machine may be used either as a dynamo or as a motor.

The effect of increasing the number of coils in series in the armature of a dynamo is to increase the E.M.F. generated.

Fluctuation of E.M.F. is prevented by spacing the coils evenly round the core.

The field magnets may be either of the 'shunt' or 'series' type as in the case of the motor, but it must be remembered that in the dynamo the armature is the seat of E.M.F., and therefore in the shunt dynamo the field circuit is in parallel with the external circuit, the whole current passing through the armature.

We have seen that the coil when acting as a motor is subject to mechanical forces forming the driving couple, and we shall show presently that when used as a dynamo the coil is also acted on by mechanical forces which form a couple resisting the rotation of the coil, and hence requiring the expenditure of mechanical work in the production of electrical energy.

11. Dual Aspect of Dynamo or Motor.

The foregoing remarks on the subject of a coil which can rotate in a magnetic field can now be reduced to two simple propositions:

1. *If such a coil is rotated by the application of an external couple an electromotive force is induced in the coil.*

2. *If a current is passed through such a coil then the coil is acted upon by a couple.*

Dynamo.

If the commutator strips of the coil in the first of these propositions are connected by an external circuit, a current flows through the coil and consequently the coil also comes under the second proposition.

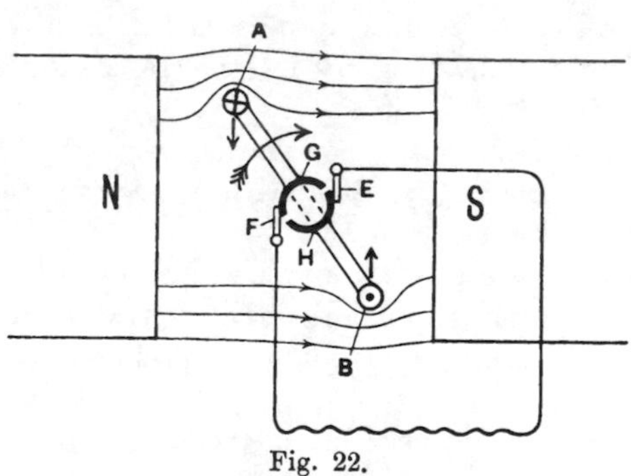

Fig. 22.

The diagram shows such a coil consisting of two conductors A and B, and *both* A *and* B must be considered as coming under both propositions (1) and (2) as stated above.

For the sake of clearness, however, we will consider A from the point of view of proposition (1) and B from the point of view of proposition (2).

It is evident from the figure that the direction of the induced E.M.F. is down A and up B, and the current will therefore flow in this direction.

Now considering the conductor B under proposition (2) we

see that a force acts on B in the direction shown by the straight arrow. An equal and opposite force of course acts on A, and thus a couple comes into action which tends to resist the rotation of the coil.

It is in overcoming this resisting or 'back' couple that mechanical work is done, and the mechanical work so done is converted into electrical energy.

In the particular case where there is no external circuit, no current flows, and there is therefore no back couple and no mechanical work is done.

However, since there is no current no electrical energy is generated.

Motor.

Referring now to proposition (2), the coil when supplied with a fixed external E.M.F. is traversed by a current and is therefore acted upon by a couple. This couple produces an acceleration in the coil, which therefore acquires an angular velocity, thereby coming under proposition (1).

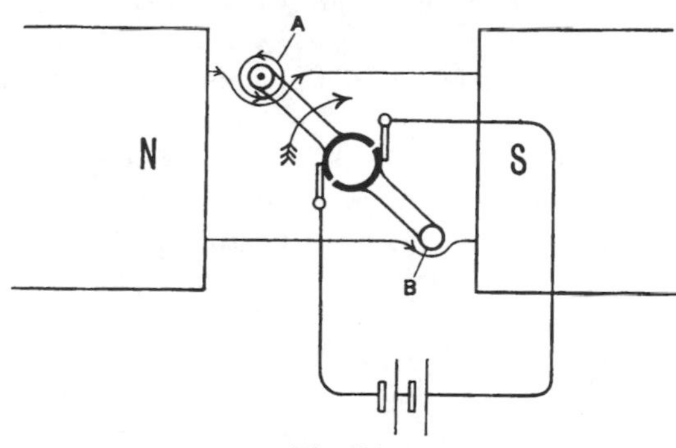

Fig. 23.

For the sake of clearness in the diagram consider the conductor A from the point of view of proposition (2) and B from the point of view of proposition (1), though of course really both conductors come under both propositions.

The direction of the current is up A and down B.

From the diagram [refer to conductor A] it is clear that the direction of rotation of the coil is clockwise.

Now consider B under proposition (1) and it is evident that an electromotive force is induced in B *tending* to send a current up B. An equal electromotive force is of course induced in A tending to send a current down A.

We see then that as soon as the coil begins to rotate an electromotive force is induced which tends to stop the flow of current. This is called the '*Back Electromotive Force.*'

The magnitude of this electromotive force is (as will be proved later) proportional to the rate of cutting lines of force, and hence if the magnetic field does not vary, it is proportional to the speed.

Broadly speaking, the actions take place as follows:

The coil is at rest to begin with.

On starting the current, a driving couple is set up and the coil receives an acceleration.

As soon as the coil begins to move an E.M.F. is generated which has the effect of diminishing the current.

The diminution of the current entails a diminution of the driving couple and therefore of the acceleration, *but not of the speed.*

The speed continues to increase and with it the back E.M.F., the current diminishing until the driving couple is reduced so far as to be just equal to the resisting couple due to friction or mechanical work which the motor is doing. When this state of affairs is reached, the 'unbalanced' couple on the coil becomes zero, the acceleration becomes zero and the speed becomes constant.

CHAPTER III

BEHAVIOUR OF PRACTICAL FORMS OF MACHINES

12. Winding of Field Magnets of Motors and Dynamos.

The field magnets may be either permanent magnets or electro-magnets. The former are not used in continuous current machines.

The latter are divided into four classes according to the nature of the circuit which carries the exciting current.

1. Separately excited.
2. Series wound.
3. Shunt wound.
4. Compound wound.

(1) *Separately excited.*

The field-magnet circuit is connected to a source of E. M. F. entirely separate from that which supplies current to the armature

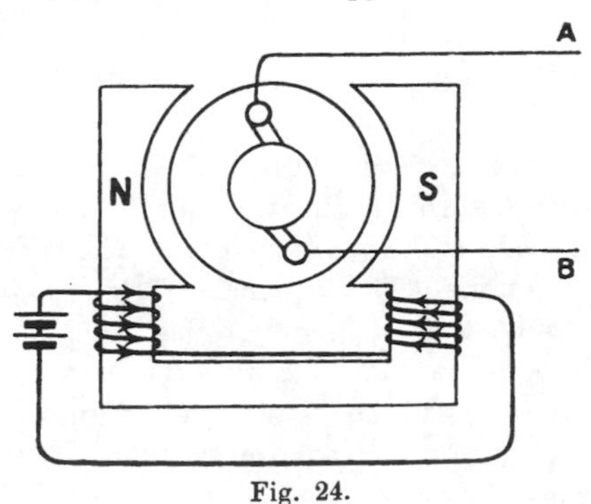

Fig. 24.

or main circuit, and there is no electrical connection between the armature and field-magnet circuits.

A and B are connected to an external circuit or to a source of E.M.F. according as the machine is to be used as a dynamo or as a motor. In the former case the armature itself is the seat of E.M.F.

The strength of the magnetic field is practically independent of the strength of the current in the armature circuit.

(2) *Series wound.*

The field-magnet circuit is connected in series with the armature, and the same current flows through both circuits.

Since the field-magnet circuit carries the whole current, the requisite number of 'ampère turns' (see § 2), can be obtained with comparatively few turns of wire, and for the same reason the wire must be of low resistance and therefore thick.

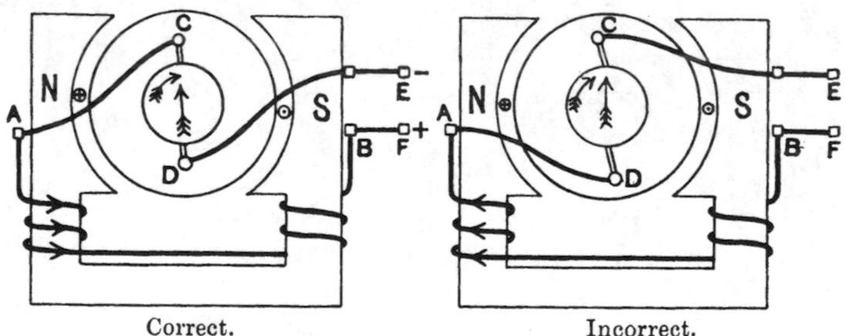

Correct. Incorrect.
Fig. 25. Series Dynamo.

Suppose the armature is so constructed that the current in passing from D to C through the armature flows down conductors on the left-hand side and up conductors on the right-hand side.

Let E and F represent the terminals of the external circuit.

We will consider the machine first as a *dynamo*.

There must be some residual magnetism in the field magnets, otherwise no E.M.F. will be generated. Suppose the residual magnetism is 'N' in the left pole piece, then if the armature is rotated clockwise the E.M.F. generated in the conductor on the left-hand side is downwards, and if the circuit is completed current flows out of C.

If C is connected to A as in the diagram the current passes round the field magnets in such a way as to reinforce the residual magnetism, and the strength of the field, induced E.M.F. and current increase together.

If the connections are reversed, that is if C is connected to E and D to A, the initial current flows round the field magnets in such a way as to destroy the residual magnetism and the dynamo will fail to generate any E.M.F.

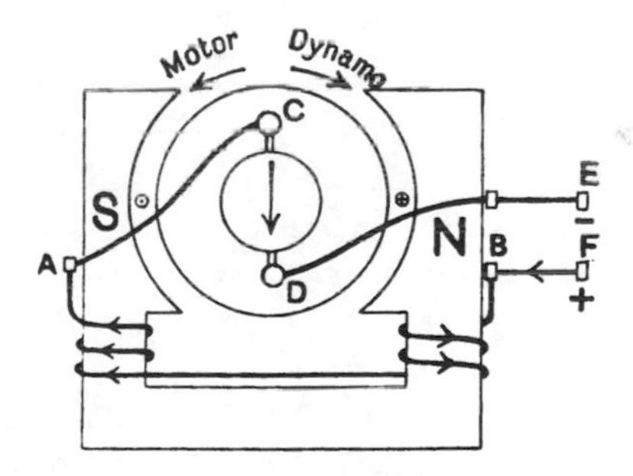

Fig. 26. Series Motor.

Now take it as a motor with the same connections, as shown in Fig. 26; if a battery is connected with its positive pole to F and its negative pole to E, current flows from C to D down all conductors on the left side of the armature and also makes the left pole 'north'; the direction of rotation is therefore anti-clockwise.

If the connections of the battery at E and F are reversed, the direction of the current in the armature *and* the polarity of the field magnets are both reversed in consequence. The direction of rotation of the armature is therefore not changed.

To reverse the direction of rotation of the motor the direction of current must be reversed *either* in the armature *or* in the field-magnet circuit, but not in both.

That is, either E must be connected to C and D to A, as in Fig. 27, or C must be connected to B and F to A as in Fig. 28.

The former method is the one employed in practice.

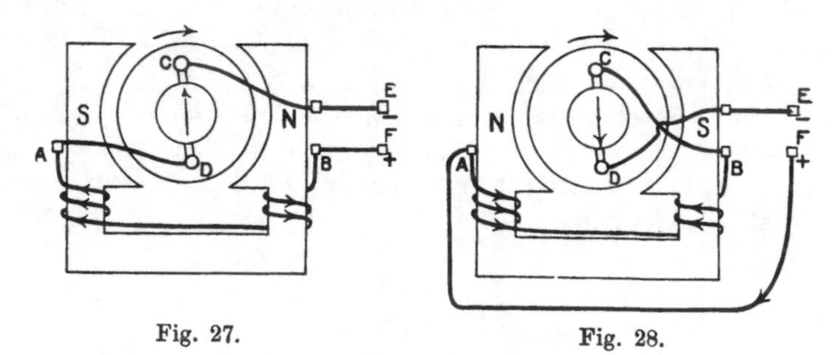

Fig. 27. Fig. 28.

(3) *Shunt wound.*

The terminals of the field-magnet circuit are connected to the brushes so that the field-magnet circuit forms a shunt to the external circuit in the case of a dynamo and a shunt to the armature circuit in the case of a motor.

Only a small fraction of the total current must be taken by the field-magnet circuit, so that to produce the requisite strength of field a large number of turns of fine wire are used.

The remarks made on the excitation of series dynamos apply also to shunt dynamos, substituting Fig. 29 for Fig. 25.

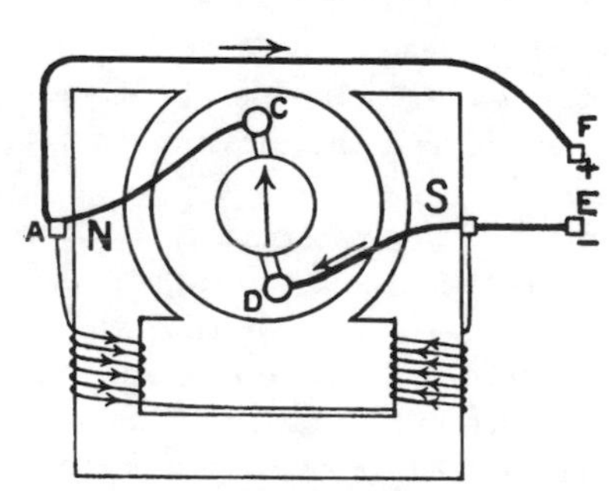

Fig. 29. Shunt Dynamo.

If the machine is to be used as a motor and if with the same connections as shown in Fig. 29 a battery is connected with its + pole to F and its – pole to E, the polarity of the field magnets remains unchanged while the armature current is reversed.

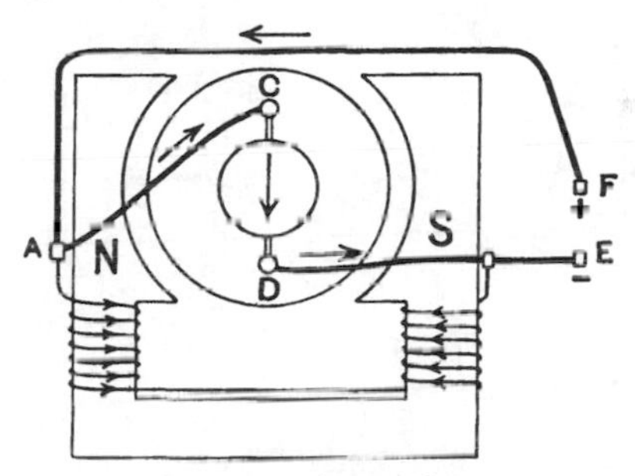

Fig. 30. Shunt Motor.

It follows that the couple generated by the armature is reversed, and since this is the driving couple in the motor whereas it was a resisting couple in the dynamo, the direction of rotation is the same in both cases.

It should be remembered that the series machine treated in this manner differed from the shunt machine in that the directions of rotation of the motor and dynamo were opposite to one another.

To reverse the direction of rotation of the motor, the direction of the current must be reversed *either* in the armature *or* in the field-magnet circuit, but not in both.

(4) *Compound wound.*

There are two circuits round the field magnets, one being in series with the armature and the other a shunt connected to the brushes.

The great majority of 'ampère turns' are due to the shunt circuit, the number of turns in the series circuit being very small.

It may be remarked that if the winding is such that when the machine is used as a dynamo the current flows in the same direction round both the field circuits, then when the machine is used as a motor the current will flow in opposite directions round the two field circuits.

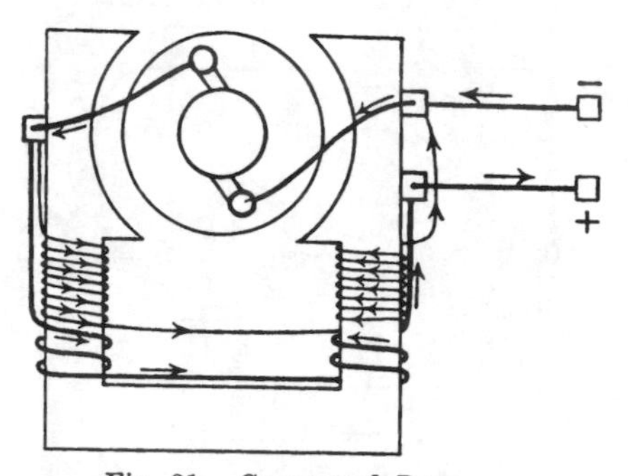

Fig. 31. Compound Dynamo.

13. Behaviour of Shunt, Series and Compound Wound Machines.

This subject will be dealt with more fully later on and for the present we will confine ourselves to brief statements which may be verified by experiment.

A shunt wound motor runs at fairly constant speed for all reasonable loads, the speed falling very slightly as the load is increased. It is therefore used for driving machines. If it be desired to vary the speed without altering the D.P. of the mains, a variable resistance is put in the field circuit; the speed increases when this resistance is increased. The range of speeds under full load is not large, however, and if this is needed, a series motor must be used.

When starting a shunt motor, the following points must be remembered.

(1) The resistance of the armature is small, so that unless the armature is generating a considerable back E.M.F. the full pressure of the mains would send a dangerously large current through it. *A resistance must therefore be connected in series*

with the armature, and this must be cut out step by step as the armature gains speed and the back E.M.F. *increases.*

A shunt motor should be allowed to attain full speed before it is loaded.

(2) In order to create as large a starting torque as possible, the field magnets should be excited to their full strength before the armature circuit is completed; *the field circuit must therefore be connected direct to the mains without additional resistance before, or at least as soon as, the armature circuit is completed.*

These operations are effected by a 'starting switch' as shown in Fig. 32.

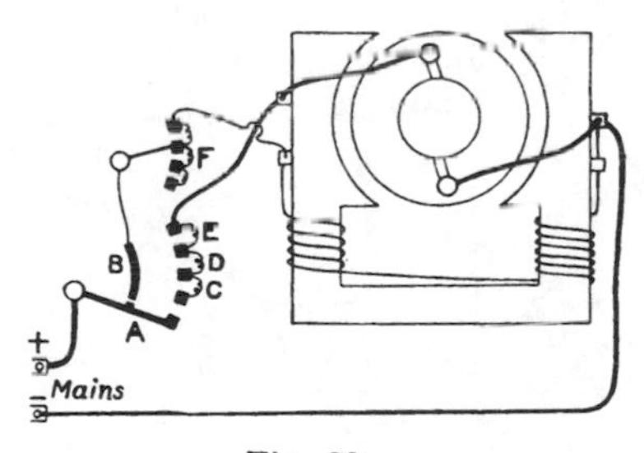

Fig. 32.

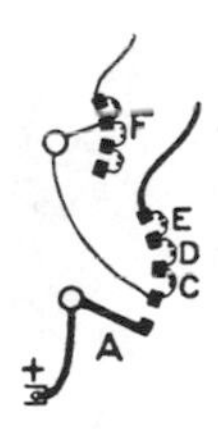

Fig. 33.

A is an arm which makes contact with B and excites the field magnets; it then connects with a stud which passes current to the armature through resistances C, D, etc., which are cut out by further motion of A, the contact with B being maintained.

For small machines the sliding contact B is sometimes omitted, as in Fig. 33; the starting resistances are then left in circuit with the field magnets when the machine is running, but it should be noted that when the arm A is on the first stop there is no additional resistance in the field circuit.

F is the speed regulating resistance, which must be all 'out' when starting.

A series motor has a wide range of speed; it can be started from rest under full load. It is therefore used for traction, cranes, boat hoists, etc. At starting a resistance must be interposed, which is cut out as the speed increases, so as to maintain the torque constant in spite of the tendency of the increasing back E.M.F. to reduce the current taken from the mains. At very light loads a series motor tends to 'race.'

Both forms of motor should be provided with an 'overload release,' which will break circuit if the load is in danger of pulling up the motor. This is shown in Fig. 34; an iron

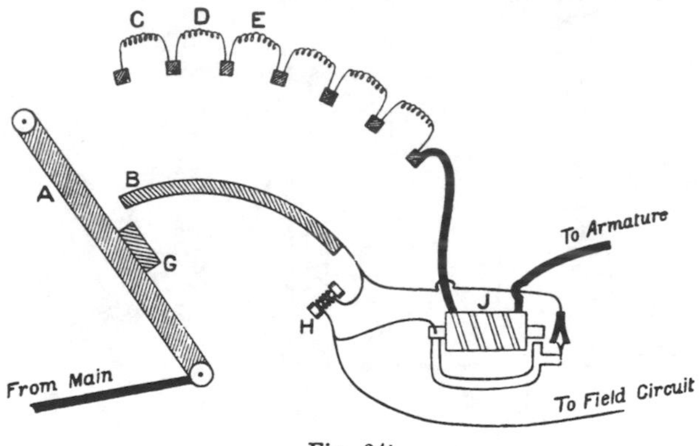

Fig. 34.

armature G is fixed to the arm A, which when A reaches the end of its travel is held by an electromagnet (H) excited by the current in the field circuit. If this electromagnet releases the armature, A flies back under the influence of a spring, so stopping the current.

Another small electromagnet (J) is provided, excited by the current passing through the armature; if this current is excessive, the second electromagnet attracts its armature and so short circuits the first, thus releasing the arm A.

A series dynamo gives practically no E.M.F. on open circuit. As the external resistance is cut out the E.M.F. increases almost in proportion to the current at first and very rapidly. As the field magnets become saturated the E. M. F. tends to a maximum value and though this remains fairly constant for any further increase of current, the terminal D. P. begins to diminish owing to the increase of 'lost volts' in the armature.

A shunt wound dynamo gives nearly constant E.M.F. at constant speed, the E.M.F. falling slightly as the current is increased, being maximum on open circuit.

If a very constant E.M.F. is required, a compound wound dynamo is used; the few 'series' coils counterbalance the falling off in E.M.F. of the shunt dynamo as more current is taken. This machine is used for electric lighting.

CHAPTER IV

ENERGY LOSSES AND EFFICIENCY

14. Energy Losses.

In every electric machine, whether dynamo or motor, there will be certain losses of energy, that is to say the energy obtained from the machine will always be less than that which is supplied to the machine. All the lost energy will ultimately appear in the form of heat.

The losses may be accounted for as follows :

1. Mechanical losses.
2. Energy lost in overcoming the resistance of the armature.
3. Energy lost in overcoming the resistance of the field-magnet circuit.
4. Loss due to eddy currents set up in the armature core and pole pieces.
5. Loss due to hysteresis.

15. Mechanical Losses.

These are due to friction in the bearings, between the brushes and commutator, etc., and are overcome as far as possible by good workmanship in the construction, suitable lubrication and so on.

16. Loss due to Armature Resistance.

Suppose we are taking a current of C ampères from the armature and that the resistance of the armature is R ohms. Then due to resistance there there will be a drop in voltage along the armature of CR volts.

The power lost in driving the current through the armature, whether the armature is used as a dynamo or as a motor, is C^2R watts.

To prevent these losses of E.M.F. and of power the resistance of the armature must be made as small as possible.

17. Loss due to Resistance of Field-magnet Circuit.

Here again the power lost is $C_m{}^2 R_m$ where C_m is the current in the field circuit and R_m the resistance of this circuit.

Other things being equal, the nature of the winding of the field magnets does not affect the expenditure of power used in exciting them. (The question whether a shunt series or compound winding is to be used is decided by considering the purpose for which the machine is required.)

For example—

Let r = mean radius of the field coils.
d = diameter of wire.
s = specific resistance of wire.
R_m = resistance of field circuit.
C_m = current in field circuit.
A_t = number of ampère turns $= nC_m$.
n = number of turns.

Then
$$R_m = 2n\pi r \cdot \frac{4}{\pi d^2} \cdot s.$$

$$\text{Power lost} = \frac{C_m{}^2 \cdot 8 \cdot n \cdot r \cdot s}{d^2} = 8 \cdot A_t \cdot \frac{C_m}{d^2} \cdot r \cdot s.$$

If the current density is the same in every case, $\frac{C_m}{d^2}$ is a constant, and hence the power used is directly proportional to the number of 'ampère turns,' and is therefore independent of the current strength provided the product of the current and number of turns is constant.

18. Eddy Currents.

If we suspend a metal cylinder so that it can be set rotating freely, and when it is rotating, bring it into a strong magnetic field the cylinder will come to rest almost immediately and the whole of its kinetic energy will be dissipated in the form of heat, just as though a mechanical resistance to rotation had been introduced.

It follows from this that a continuous supply of energy is necessary to keep such a cylinder rotating in a magnetic field, over and above the energy which is spent in overcoming friction.

The diagram represents one-half of such a metal cylinder rotating between the poles of an electro-magnet.

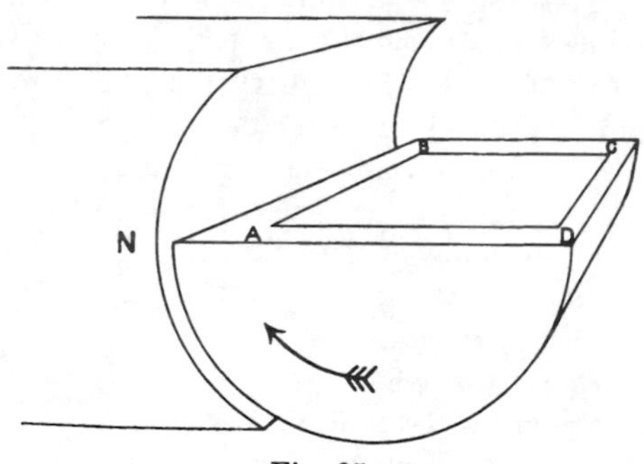

Fig. 35.

A strip of metal such as AB cuts the magnetic flux, and therefore has an E.M.F. induced in itself just as much as if it were an insulated conductor, and so also does a strip of metal such as CD. Strips such as BC and AD have no E.M.F. induced in them in the direction of their length. A current therefore flows, roughly in the direction ABCD.

In the case of a solid metal cylinder the resistance of such a path as ABCD is very small and a considerable current may flow.

Such currents are called 'eddy' currents. Electrical energy is developed in the production of these currents at the expense of mechanical work and this energy is finally dissipated in the form of heat.

To prevent this loss of energy the armature core is made up of a number of thin stampings or laminae which are insulated with paper or varnish, the arrangement being such as to produce a large resistance to the flow of eddy currents without sensibly increasing the reluctance of the core to the passage of the magnetic flux.

The diagram shows a section of such a laminated core.

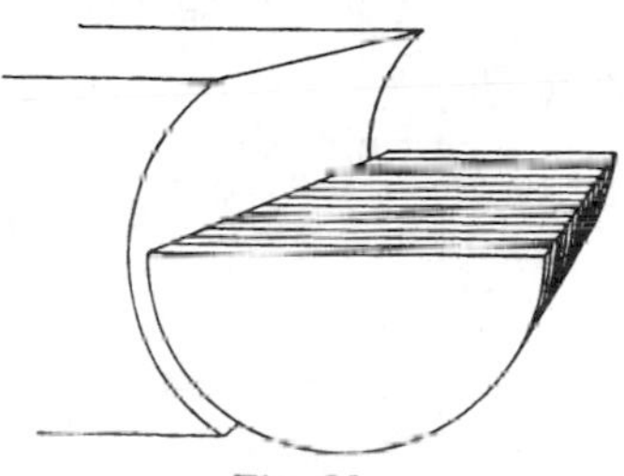

Fig. 36.

When the armature winding is placed in slots in the core it becomes necessary to laminate the pole pieces of the field magnets. For the lines of force enter and leave the armature core in 'tufts' by way of the projecting 'horns' of the core and consequently a part of the metal of the pole piece which is opposite one of the horns of the core is in a stronger field than a part which is opposite a slot. Hence every portion of the pole face is subjected to a rapidly altering magnetic field with the result that eddy currents are set up which cause a loss of power and the development of heat.

Fig. 37 illustrates the lines of force entering the armature core in 'tufts' and indicates the method of laminating the pole piece.

19. Hysteresis Loss.

It can be shown that when iron is made to pass through a complete cycle of magnetisation the work done in the process is directly proportional to the area of the hysteresis curve (see p. 7).

Now the area of this curve and therefore the amount of work done in each complete cycle of magnetisation depends on the quality of the metal which is used, being least in the case of annealed wrought iron, and greatest in the case of hard high-carbon steel.

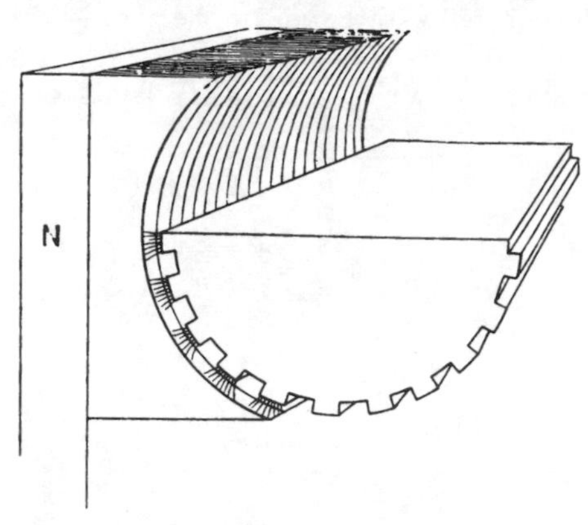

Fig. 37.

The armature core of a two pole dynamo or motor passes through a complete cycle of magnetisation once every revolution, and there is consequently a loss of power which can be diminished but not entirely avoided by the use of suitable material.

We will now show that 'hysteresis' in an armature core generates a couple which opposes motion.

Fig. 38 represents an armature core between two poles. [The pole pieces are made small for the sake of clearness in the explanation, but it will readily be seen that the same theory will hold good for large 'pole faces.']

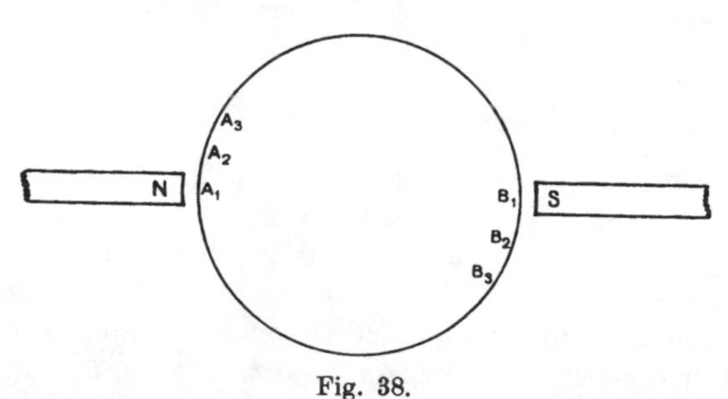

Fig. 38.

If the armature core were at rest there would be a S. pole at A_1 and a N. pole at B_1.

If it were not for the magnetising effect of the field magnets, then on rotating A_1 to A_3 the S. pole which was at A_1 would simply rotate with the core to A_3; but the field magnets tend to change the position of the poles of the core relative to the core and keep it fixed relative to the field magnets.

The hysteresis of the iron tends to keep the magnetism of the core in the direction A_3B_3, the result being that the magnetic axis of the armature core is distorted to an intermediate position A_2B_2, and this produces a 'back' couple on the armature.

A certain amount of power is used up in overcoming this back couple when the armature is rotating.

This principle is utilized in Ewing's Hysteresis Tester.

Iron for use in the armature core should have a high permeability and a small hysteresis curve.

Iron for use in the field magnets should have a high permeability, but it does not matter whether its hysteresis curve is small or not.

The qualities of high permeability and low hysteresis often go together, but should be carefully distinguished.

20. Efficiency of Dynamos.

The losses of energy which occur in the electric machine may be roughly divided into mechanical losses and electrical losses.

In treating of the various losses separately we mentioned:

(1) Frictional losses.

(2) Losses due to resistance in armature and field magnet circuits.

(3) Eddy-current losses.

(4) Hysteresis losses.

Of these, numbers (1) and (2) may be called direct losses, (1) being mechanical and (2) electrical.

Numbers (3) and (4) may be called indirect losses, and for the sake of simplicity in dealing with the efficiency of electric machines they will be included under mechanical losses.

In the case of a dynamo the formation of eddy currents in

armature core and pole pieces and the hysteresis of the armature core produce a couple which tends to prevent rotation. Due then to both eddy currents and hysteresis a greater driving torque is necessary and the indirect effect of these two phenomena is a mechanical loss.

In any dynamo the *electrical efficiency* is the ratio of the externally useful electrical power to the total electrical power generated.

The *commercial efficiency* is the ratio of the externally useful electrical power to the mechanical power used in driving the machine.

In a 'series' dynamo the electrical quantities which can be directly measured are the current and the D. P. at the terminals of the machine.

The following symbols are used:

C = current in ampères.

V = D. P. at terminals in volts.

E = total E. M. F. generated in volts.

e = fall in volts along the armature and field circuits, sometimes called 'ohmic drop.'

R_a = resistance of armature in ohms.

R_m = Resistance of field circuit in ohms.

The useful electrical power is $V \cdot C$ watts.

The total electrical power generated is $E \cdot C$ watts.

Now $$E = V + e = V + C\,(R_a + R_m).$$

Hence the electrical efficiency is given by

$$\frac{V \cdot C}{VC + C^2(R_a + R_m)}.$$

In a 'series' dynamo V is of course the D. P. between one brush and the other end of the field circuit, not the D. P. between brushes.

The commercial efficiency of a 'series' dynamo is given by

$$\frac{V \cdot C}{(\text{Mechanical H. P. used}) \times 746}.$$

Since the commercial efficiency is the product of the mechanical and electrical efficiencies, the mechanical efficiency can be obtained by dividing the commercial efficiency by the electrical efficiency.

It must be remembered that the mechanical efficiency so found takes into account both eddy currents and hysteresis losses.

In a 'shunt' dynamo the electrical quantities which can be measured directly are the current in the external circuit and the D. P. between the brushes.

The following symbols are used:

V = D. P. between brushes in volts.
C_x = current in external circuit in ampères.
C_a = current in armature in ampères.
C_m = current in field circuit in ampères.
R_m = resistance of field circuit in ohms.
R_a = resistance of armature in ohms.
e = 'ohmic drop' in armature in volts.
E = total E. M. F. generated in volts.

In this case C_a is the total current and is equal to $C_x + C_m$.

$$C_m = \frac{V}{R_m}, \quad E = V + e = V + C_a R_a,$$

hence the electrical efficiency

$$= \frac{VC_x}{EC_a} = \frac{VC_x}{\left(C_x + \frac{V}{R_m}\right) V + R_a \left(C_x + \frac{V}{R_m}\right)^2}$$

$$= \frac{VC_x}{VC_x + \frac{V^2}{R_m} + \left(C_x + \frac{V}{R_m}\right)^2 R_a}.$$

This last expression is in the form of

$$\frac{\text{Useful Power}}{\text{Useful Power} + \text{Power lost in Field Circuit} + \text{Power lost in Armature}},$$

and further, it is in terms of quantities which are directly measurable.

21. Efficiency of Motors.

The power used by a motor is given by EC in watts, where E is the E. M. F. of the main supply, C the total current taken by the motor. Of this electrical power used part is wasted in heat and the remainder used to generate mechanical power. Of the mechanical power so generated a further part is wasted in overcoming the effects of eddy currents, hysteresis in the armature core and frictional losses, the remainder being available for useful work.

Series motor.

Let R = resistance of armature and field circuit in ohms.

C = total current in ampères.

e = back E.M.F. in volts.

E = E.M.F. of Mains in volts.

Then the total power used = EC watts.

$$C = \frac{E - e}{R}.$$

Power wasted in heat $= C^2R = (E - e) . C$, hence power converted into mechanical power is $EC - (E - e)C = eC$ watts.

Hence electrical efficiency $= \frac{eC}{EC} = \frac{e}{E} = \frac{E - CR}{E}.$

Total or commercial efficiency is given by

$$\frac{\text{Brake Horse Power} \times 746}{EC}.$$

Shunt motor.

As in the case of a series motor the total power used is given by EC, where C is the total current.

The mechanical power generated is as before $e . C_a$ watts,

and $$C_a = C - \frac{E}{R_m}, \quad e = E - C_a R_a.$$

Hence the electrical efficiency is

$$\frac{eC_a}{EC} = \frac{\left[E - R_a\left(C - \frac{E}{R_m}\right)\right]\left(C - \frac{E}{R_m}\right)}{EC}.$$

The total efficiency is $\dfrac{\text{B.H.P.} \times 746}{EC}.$

These expressions for efficiencies are not intended to be used as formulæ, but are to illustrate the methods of calculation.

CHAPTER V

THEORY AS TO THE EXCHANGES OF ENERGY IN ARMATURES

22. Quantitative Meaning of Magnetic Lines of Force.

We have hitherto used lines of magnetic force as showing merely the *direction* of the resultant magnetic force at any point of the field of force, but have not settled how far apart we should draw the lines. We will now extend the definition of 'lines of force' so that they will show the intensity as well as the direction of the magnetic force at the point.

The 'intensity of the field' at any point is measured by the mechanical force in dynes acting on a *unit* north pole at that point, tending to move it in the direction of the line of force.

For simplicity, consider first a uniform field, in which all the lines of force are parallel to one another. Draw as many magnetic lines passing through each square centimetre as there would be dynes acting on a unit pole. The square centimetre is of course to be taken at right angles to the force. Therefore if the field is strong, the lines will be crowded together, if weak, they will be far apart; and the 'density' of the lines will be a measure of the strength of the field.

Next as a less simple case consider the field round an isolated magnet pole of strength m. It will consist of lines of force radiating from the pole in all directions. By the second law of magnetic force, the intensity of the field at a distance r cm. from the magnet pole is $\frac{m}{r^2}$. If then we wish to represent the force intensity at points on this sphere by the density of the lines of force in the same way as before, we must draw them so that $\frac{m}{r^2}$ lines pass through each square cm. of the surface of this sphere. Since the area of the surface of a sphere of r cm. radius is $4\pi r^2$ sq. cm., there will be $4\pi r^2 \times \frac{m}{r^2}$ or $4\pi m$ lines to be drawn in all. If we produce these same lines to another concentric sphere of radius r_1 cm., their density will change so that there are $4\pi m$ lines to $4\pi r_1{}^2$ sq. cm., i.e. $\frac{m}{r_1{}^2}$ to 1 sq. cm.; and $\frac{m}{r_1{}^2}$ is the field intensity at that distance. Thus we see that if we draw the correct number of lines per sq. cm. at one point they will contract or spread out so as to represent correctly the field intensity at all other points of their course. This can be shown to be true of all fields of force, however complicated.

Note that if we take $m = 1$ in the above, we find that *there proceed from a unit pole 4π 'lines of force.'*

Thus a field of magnetic force can be completely represented by lines of force, the direction of the force at any point being that of the lines of force there and its intensity being the number of such lines to the sq. cm.

It is to be understood that in a field of force which alters from point to point, the force at any point must not be taken as the number of lines through an actual sq. cm. round the point, but a very much smaller area containing the point must be taken and the number of lines passing through it must be found. Then the 'number per sq. cm. at that point' means the number that would pass through a sq. cm. if the density were the same all over it as it is over the small area. This is exactly analogous to the case of hydrostatic pressure at a point.

23. Absolute Electromagnetic Units.

The units adopted in practice are the Ampère, Volt and Ohm; the ampère and volt being defined by means of the relation between electric currents and chemical changes, the ohm being derived from them.

These appear to be very arbitrary units, as it is not obvious why the numbers ·001118 and 1·434 occurring in their definitions should have been selected. But the real basis for the units is a theoretical one depending on the centimetre, gramme and second only, and from these the practical units are derived.

In the same way the metric unit of length is defined as one ten millionth of the distance of the pole to the Equator, but owing to the impossibility of applying this in everyday practice, the legal unit of length is defined to be the distance between marks on a standard bar which was made as nearly as possible of this length; and if we defined a foot in terms of this length we should get a number apparently as arbitrary as that occurring in the definition of the ampère.

This theoretical system is based on the relations which hold between electric currents and lines of magnetic force. It has been shown on page 8 that when a current-bearing conductor is situated in a field of magnetic force, the conductor is acted on by a mechanical force which depends on the intensity of the magnetic field.

Current.

We may define our unit of electric current by means of this fact, as follows:

Suppose that a conductor carrying a current is situated in a field of magnetic force of unit intensity, so that it is at all points of its length at right angles to the lines of force which cut it at those points; then if it is acted on by a mechanical force of 1 *dyne per centimetre, the conductor is said to carry unit electric current.*

It will be seen that we have here a definition of unit current which depends solely on the gramme, centimetre and second and the fundamental properties of magnet poles and electric currents; it is therefore called an *Absolute Electromagnetic Unit.*

We have only stated that a unit current in a unit field experiences unit mechanical force per cm. of its length; we will extend the definition and say that under the same conditions if there is a force of F dynes per cm. of the conductor when it is in a magnetic field of intensity H lines per sq. cm., there is a current of $\frac{F}{H}$ absolute units flowing in it.

This unit is inconveniently large for practical use, so that it was decided to adopt a unit one-tenth as large; this was called an *ampère*, and after much experimental research the value ·001118 gm. was agreed upon as best representing the amount of silver deposited per second by a current of one-tenth of the absolute electromagnetic unit. This practical unit was adopted as the legal standard by an Order in Council of August 23rd, 1894.

***Reduction Factor of Tangent Galvanometer.**

We can see how this definition leads to the value of the reduction factor of a tangent galvanometer.

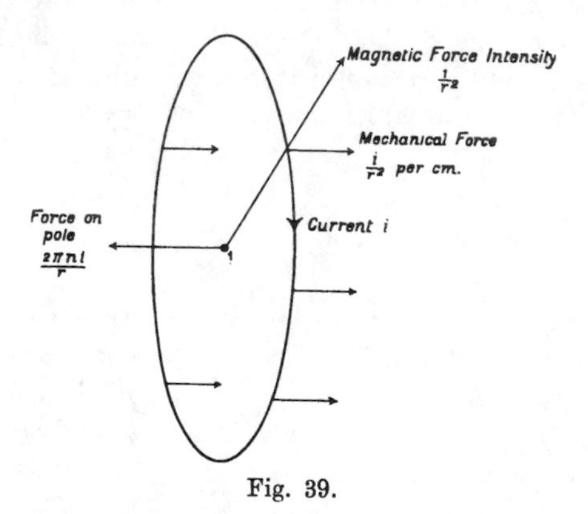

Fig. 39.

Suppose that i electromagnetic units of current are flowing round a circular coil of wire of radius r cm. and n turns, and that a unit magnetic pole is situated at the centre of the ring. This unit pole will produce a magnetic field of intensity $\frac{1}{r^2}$ lines per sq. cm. at all points of the ring, the lines of force being at right angles to the conductor, so that the mechanical force on each cm. of the wire is $\frac{i}{r^2}$ dynes. Now the coil has a length of $2\pi rn$ cm. and all the forces on the wire are parallel, so that the total force on the coil is

$$2\pi rn \frac{i}{r^2} \text{ or } \frac{2\pi ni}{r} \text{ dynes.}$$

But by Newton's third law of motion, the unit magnetic pole must itself be acted on by this force; let us call it F, then

$$F = \frac{2\pi ni}{r} \text{ dynes.}$$

If we express the current in ampères as C, then $C = 10i$, since each ampère is $\frac{1}{10}$th of an absolute unit;

$$\therefore \quad F = \frac{2\pi nC}{10r} \text{ dynes.}$$

This leads to the reduction factor $\frac{10Hr}{2\pi n}$. Accurate experiments with such a galvanometer in series with a silver voltameter have led to the silver depositing power of a current which is adopted as the legal standard.

Quantity.

The electromagnetic unit quantity of electricity is directly derived from this, as *the quantity conveyed by unit current in one second*; the practical unit is the quantity conveyed by an ampère in one second, and is called a *Coulomb.*

E.M.F.

The absolute electromagnetic unit of E.M.F. is defined as follows:

Two points on a conductor are said to have unit E.M.F. *between them when one erg of work is done by the electric forces in urging unit quantity of electricity from one point to the other.*

For a practical unit of E.M.F. it is found convenient to take *one hundred million* of these absolute units—it is called a *Volt.* Experiments show that the E.M.F. of a Clark cell, under certain standard conditions, may be best expressed as 1·434 of these volts; and this has been adopted as giving the legal definition of the volt.

Resistance.

The value of the unit of resistance should be chosen so that no numerical coefficient may be required in Ohm's Law; hence the value of the absolute unit of resistance depends on those of current and E.M.F., and the ohm depends on the ampère and volt. The relation between the ohm and the absolute unit of resistance is found thus:

$$1 \text{ ohm} = \frac{1 \text{ volt}}{1 \text{ ampère}} = \frac{10^8 \text{ absolute units}}{10^{-1} \text{ absolute units}}$$
$$= 10^9 \text{ absolute units.}$$

The legal standard ohm is defined as the resistance of a uniform 'wire' of mercury of certain dimensions.

24. Work done when a current-bearing Conductor moves across lines of magnetic force.

In order to fix ideas, imagine a bar AB of length l cm. capable of sliding along two rails CB, DA, which are connected

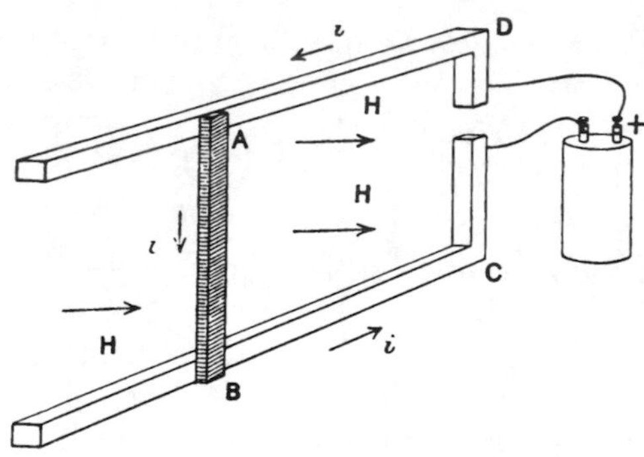

Fig. 40.

together at one end, the whole being in a plane perpendicular to a field of magnetic force of intensity H lines per sq. cm.; and imagine that there is a battery in CD which maintains a current of i absolute units in AB from A to B. AB is assumed to be at right angles to BC or AD.

Then from the definition of electric current, there is a mechanical force acting on AB of magnitude Hli dynes and in the figure this will urge AB to the left.

If now we force AB to move to the right (i.e. in a direction opposing the mechanical force exerted by the magnetic field) through a distance L cm., the work W expended in producing the motion will be $HliL$ ergs

or $$W = HliL \text{ ergs.}$$

Now during the motion it will move across HlL lines of force; represent this number by N.

Then we have $$W = iN \text{ ergs}$$

or, in words,

The work done is the product of the current in absolute units into the total number of lines of force cut by the conductor.

This result holds true whatever shape the current-bearing conductor may have.

It is important to note that if the conductor is moved against the mechanical force, produced by the magnetic field on the current-bearing conductor, work must be done and energy must be expended by the agent which produces the motion; but if the conductor is allowed to move in the direction of the force, work will be done by the moving conductor.

We will next consider where this energy goes to in the first case, and where it is drawn from in the second.

25. Energy required to maintain a Current.

The definition of absolute unit E.M.F. between two points of a conductor states that its amount is expressed by the work done by the electric forces in the movement of a unit quantity of electricity along the conductor. In considering the 'E.M.F. round a complete circuit,' we must modify this definition by saying that it is measured by the work which is done by the electric forces in order to cause a unit quantity of electricity to move *once* round the circuit. But a current expressed by i absolute units is equivalent to the passage of i units of electricity once round the circuit per second. We see then that if there is an E.M.F. of E absolute units round a circuit, and if a steady current of i absolute units flows round the circuit, work is done by the electric forces at the rate of Ei ergs per sec.

If we imagine our circuit to have no resistance or other means of getting rid of energy, a current once started would continue to flow undiminished, and no E.M.F. would be required to maintain it. But in the case of real circuits there is resistance; a current once started dies away almost instantly, its energy being transformed into heat, and if we wish to maintain a steady current in such a circuit we must maintain a steady E.M.F.

In order then to maintain an E.M.F. (E) *round a circuit carrying a current* (i) *we must supply energy from outside at the rate of* Ei *ergs per sec.*

26. Value of the E.M.F. produced in a Conductor which moves across lines of force in a magnetic field of given intensity.

We know that one way of maintaining an E.M.F. round a closed circuit is to use a battery; the energy necessary to keep the current flowing is then supplied by the combination of the chemicals which it contains.

But there is another way of maintaining an E.M.F. The experiments described in Art. 8 show that when a part of a circuit is made to move across lines of magnetic force, a current is induced in the circuit. To produce and maintain this current an E.M.F. round the circuit is needed; therefore we may say that the motion maintains an E.M.F. Suppose that this E.M.F. maintains a current i absolute units in the circuit we are considering.

Again in Art. 24 we have proved that mechanical work must be done to make the conductor move across the lines of force, if there is a current flowing round the circuit of which the conductor forms part; and if this current is i absolute units, and if the number of lines of force cut per second is N, then the mechanical work done per sec. is iN ergs.

This quantity of energy (iN ergs) put into the circuit each second is the energy which keeps the current flowing.

Let us denote by E the E.M.F. required to maintain a current i in this particular circuit. We have seen that if a current of i

units flows in a circuit round which the E.M.F. is E units, then the energy required per second is Ei ergs (see Art. 25).

Therefore $$iN = Ei,$$

or $$E = N.$$

In words:

When the number of lines of magnetic force passing through a conducting circuit is changed, an E.M.F. *is induced round the circuit, which is equal when expressed in absolute units to the rate of change of the number of lines of magnetic force.*

It should be noted that in order to determine the magnitude of the E.M.F. induced by the motion we have employed the current induced in the circuit by the motion; the magnitude of this current depends on the resistance of the circuit, and as we have not had to specify this it is clear that the proof holds good whatever value this current may have. This is also clear from the fact that i finally cancels out from the equation.

The induced E.M.F. therefore depends solely on the rate at which the lines of force are cut, and does not depend at all on the current flowing in the circuit during the motion. On the other hand, the mechanical work done in moving the conductor does depend on the current flowing.

An important illustration of this is the case of a dynamo driven by an external source of power such as a steam engine. When its armature is made to revolve in the magnetic field of the pole pieces, an E.M.F. is produced in it which is proportional to the rate at which the conductors in it cut the lines of force. If the circuit is closed through some resistance, this E.M.F. produces a current in the armature and the external circuit, and the work done per second in forcing the armature to turn is proportional to this current (assuming the rate of revolution of the armature to remain constant).

This work re-appears in whatever work the current generated is made to perform. Suppose that this dynamo is coupled up to a motor. When the current flows through the armature of the latter it will only expend energy there in the form of heat unless the armature is allowed to move so as to cut across lines of force. If this is the case, a 'back E.M.F.' will be generated, tending to check the flow of electricity; the forward E.M.F. of the dynamo will however force a reduced current to flow in spite of this, and we shall have a current-bearing conductor crossing lines of force. In this case it will be moving with, instead of against, the mechanical force exerted on it by the field of force, and so will be able to do external mechanical work.

CHAPTER VI

TYPES OF ARMATURES

27. Armatures.

We have described a very simple type of armature which is suitable for either dynamo or motor and we have seen that such a machine generates a mechanical couple *and* an electromotive force whenever it is used for *either* purpose. Whether the machine acts as a dynamo or as a motor depends simply on the kind of energy with which it is supplied.

Such an armature, though simple and easy to understand, is never used for practical purposes, but it is of great importance, because every continuous current motor or dynamo consists of an armature which rotates in a magnetic field and the armature itself consists of a number of coils like the one we have described connected either in series or in parallel or both.

We will now describe very briefly a few types of armature which are used in practice.

28. Shuttle Armature.

This is merely a modification of the single coil armature. It consists of a large number of coils in series with one another connected to one pair of commutator strips. In this and in all other types of armature the core on which the conductors are wound is made of iron in order to facilitate the creation of a strong field of force.

When this armature is used as a dynamo the E. M. F. generated at any moment will be approximately the same in all coils, so that although the total E. M. F. at any moment will be greater than that in a single coil armature of the same size in proportion to the number of coils in the armature, still the fluctuation of the E. M. F. will be just as great in proportion as in a single coil armature.

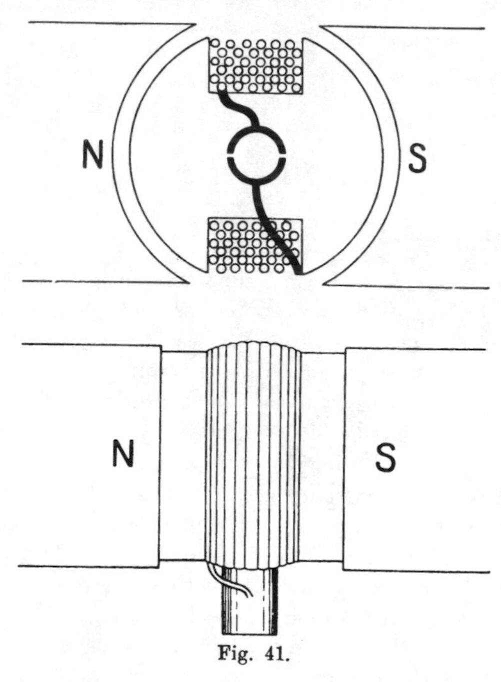

Fig. 41.

When the armature is used as a motor the driving couple will be proportionately greater at any moment than in a single coil armature, but there will be just as much fluctuation in the value of the couple as the armature rotates. This type of armature is now obsolete.

29. Ring Armature.

In this type of armature the conductors are wound on a ring of iron, some of the conductors passing inside the ring and some outside the ring.

We will consider first the state of the magnetic field near the armature core. The simplest way to get an idea of such a field is to place an iron ring between two poles of a strong magnet and explore the field with iron filings.

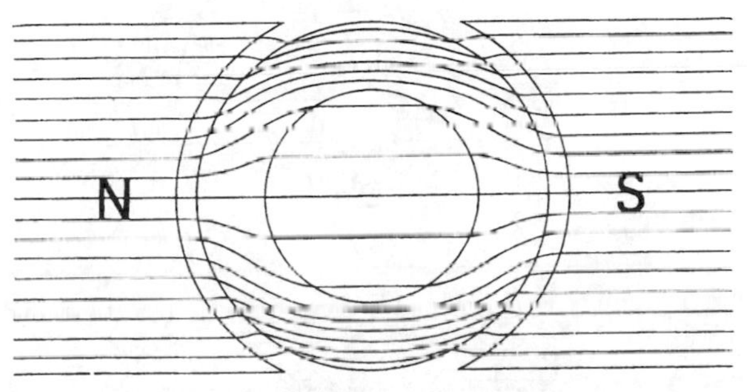

Fig. 42.

The diagram gives an idea of the nature of the field.

The lines of force which stream across from the N to the S pole of the magnet nearly all pass through the iron and only a very few bridge across the air space inside the ring, so that any conductor which is on the outside of the ring cuts practically all the lines of force, while any conductor which is on the inside of the ring is never in any more than a very weak field and may therefore by comparison be considered inoperative as far as generating an electromotive force or a mechanical couple is concerned.

30. Single Coil Ring Armature.

In this armature there is only one active conductor, AB (fig. 43).

If we assume the intensity of the magnetic field between the pole piece and the armature core to be uniform and equal to H lines per sq. cm., the lines being parallel to one another, then when this armature is used as a dynamo the E.M.F. at any instant will be $H \,.\, l \,.\, v \times 10^{-8}$ volts, where l is the length of AB in cms. and v is the linear velocity of AB *at right angles to* the direction of the lines of force, since it cuts lines at the rate of Hlv lines per sec.

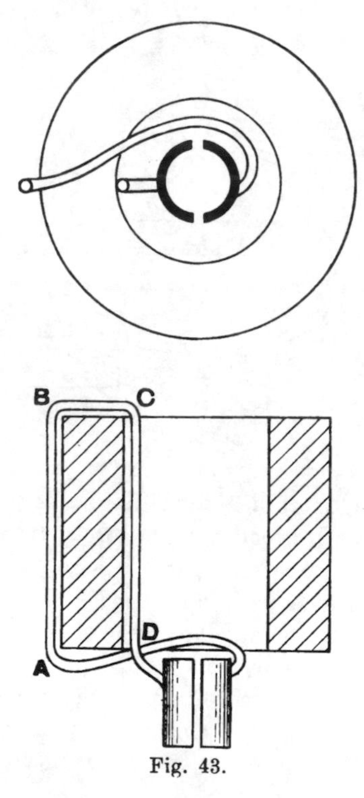

Fig. 43.

If V is the actual linear velocity of AB at any instant, then $v = V \sin \theta$ where θ is the angle turned through from the position midway between the poles of the magnet.

The curve of E.M.F. to a time base will under these circumstances be a sine curve.

31. Two Coil Ring Armature—Coils in Parallel.

In this case the two coils must generate the same E.M.F. at any instant and must therefore be diametrically opposite one another.

Considered as a dynamo the E.M.F. is the same in this case as if there were only one coil, but the internal resistance of the armature is reduced by one half.

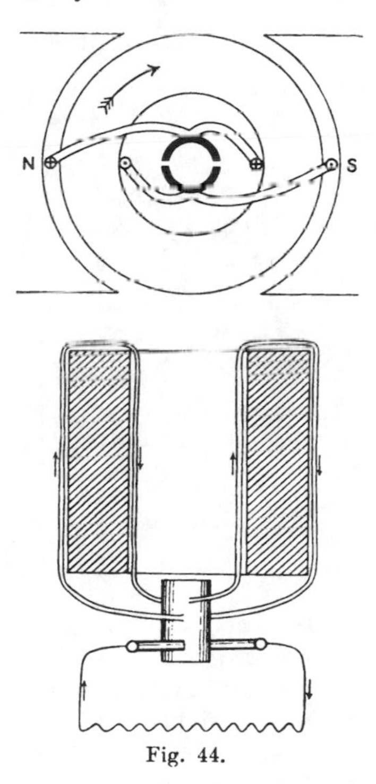

Fig. 44.

If the brushes are connected to an external circuit, half the total current flows through each coil.

The two coils are analogous to two voltaic cells connected in parallel.

32. Four Coil Ring Armature—Coils in Series and Parallel.

The coils are so arranged that one pair is on a diameter at right angles to that on which the other is situated.

There are now four commutator segments and still two brushes.

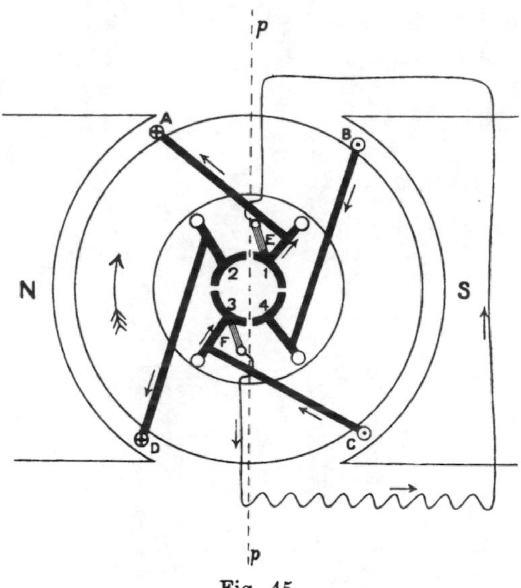

Fig. 45.

A diagram in elevation only is given.

There are here four active conductors A, B, C and D.

The E. M. F. induced in A is always equal in magnitude to that in C at the same instant, and similarly with B and D. When the E. M. F. in A and C is a maximum, that in B and D is zero. A will always be electrically in parallel with C and in series with whichever of the two B and D is cutting the magnetic field in the same direction as A. At the instant for which the diagram is drawn A and D are in series and B and C are in series, A and D together being in parallel with B and C.

The path of the current at the instant is shown by arrows.

As soon as A arrives at the plane *pp* the E. M. F. induced in it will change its direction and the position of the brushes must therefore be such that the commutator strip '2' comes under the brush E when A is in the plane *pp*. For the next half-revolution A will be in series with B and D with C.

Since A is always in series with either D or B, and the E.M.F. in D equals that in B, the total E.M.F. at any instant must equal the sum of the E.M.F.'s in A and B at that instant.

The shape of the E.M.F.-time curve will therefore be obtained by adding together the ordinates of two sine curves which are 90° apart.

Assuming that the armature rotates at n revolutions per second, the time taken to complete one revolution is $\frac{1}{n}$ seconds.

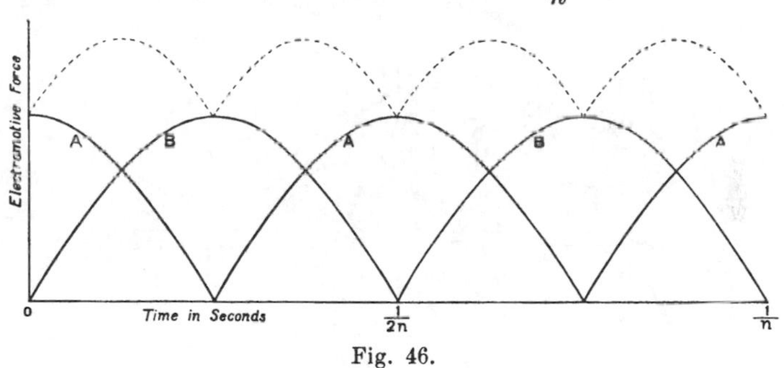

Fig. 46.

The dotted curve is that for the total E.M.F., the two full curves being the E.M.F.-time curves of A and B respectively. The average E.M.F. for the whole armature is greater than that for one coil, but the percentage fluctuation of E.M.F. is very much less.

For one coil the fluctuation is from 0 to 100, which is 100 °/。 of maximum E.M.F.

For the whole armature the fluctuation is from 100 to 141·4, or 29·2 °/。 of the maximum E.M.F.

The resistance of the armature is clearly equal to that of one coil alone. The armature is at any instant analogous to four cells, two in series and two in parallel.

There is no limit to the number of coils which can thus be wound on a ring, and the greater the number of coils the less will the fluctuation of E.M.F. be per cent. of its maximum value.

An illustration is given of an armature consisting of four

pairs of such coils and an electromotive force-time curve for such an armature is shown on the assumption that the magnetic field is uniform and that therefore the E. M. F.-time curve for one coil is a sine curve.

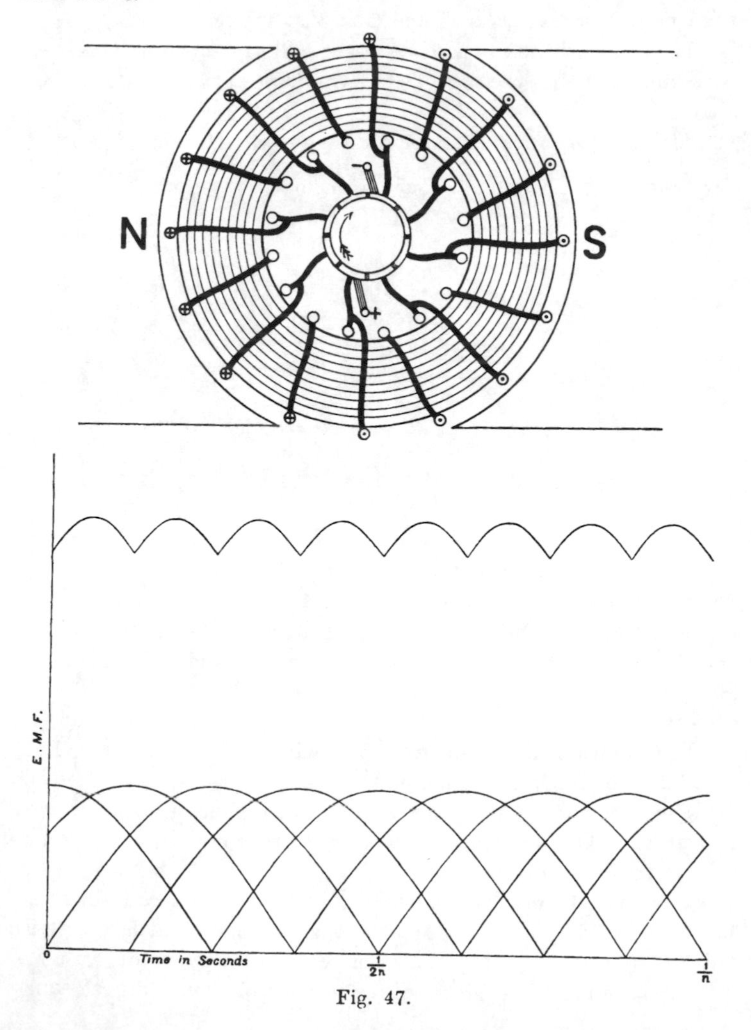

Fig. 47.

A 'ring' armature in practice will be made on the same principle as this, but will consist generally of a much larger number of coils. Each coil may consist of one or more turns.

Two advantages which this type of armature possesses over the 'drum' type which we shall consider next are:

1. No coil need overlap any other, and hence any coil can be repaired without interfering with the others.

2. The D.P. between adjacent conductors is never more than the maximum electromotive force of any one coil.

Against these advantages there is the disadvantage that there is a large number of conductors which are inoperative except as regards increasing the weight and resistance of the armature.

33. Drum Armature (Two Pole).

In a 'drum' armature all the conductors are laid along the outside of the armature core and hence with the exception of the end connections all the conductors are 'active.'

Each conductor is connected at one end to a commutator strip and at the other end to another conductor placed nearly diametrically opposite.

Like the 'ring' armature the 'drum' armature contains two paths in parallel from brush to brush. It is therefore necessary that the total E.M.F. generated in each path shall be the same at every instant, and this condition will be satisfied if one conductor of each diametral pair is in each path and the arrangement is symmetrical.

Reference to the diagram (which is of a drum armature with sixteen conductors) will make the system of connections clear.

The armature is considered as a dynamo and is shown in elevation from the commutator end. Front end connections are drawn in full, back end connections dotted. The small circles numbered 1 to 16 represent the active conductors in section. The dots and crosses show the directions of the induced E.M.F. in each case.

Beginning with conductor No. 1, connect the front end to any commutator strip, say '*a*,' and from '*a*' connect to either No. 8

or No. 10. Connection must *not* be made to No. 9, for this would put Nos. 1 and 9 in series, and as we have seen they must always be in parallel. It is necessary that the conductors to which No. 1 is connected shall 'change sides' as nearly as possible simultaneously with No. 1, hence Nos. 8 and 10 are chosen in preference to any others.

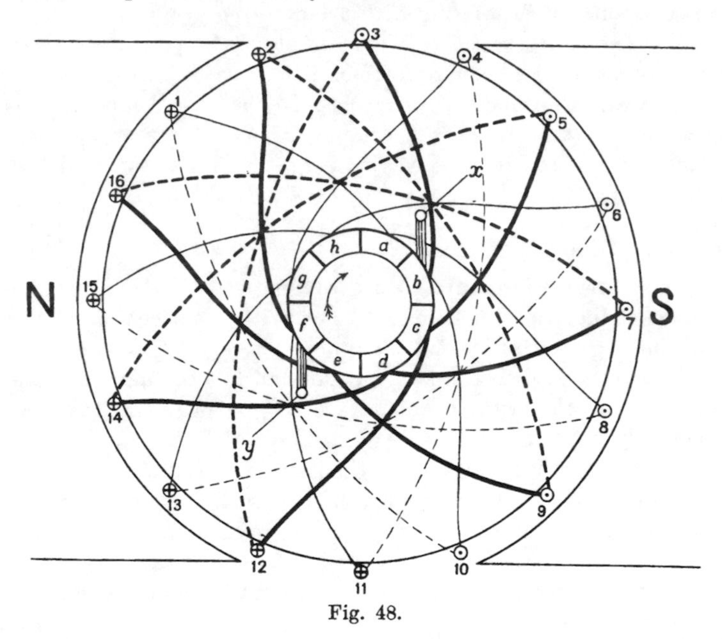

Fig. 48.

If the connection is made from No. 1 to No. 8 in front it will be from No. 1 to No. 10 at back and vice versa.

Since No. 8 is connected to No. 1 in front it must be connected to No. 15 at back and so on.

The next step is to decide on the position of the brushes.

Conductor No. 3 in the diagram is shown on the point of 'changing sides,' that is to say it was cutting the magnetic field in one direction and is just about to cut the field in the other direction, and consequently the direction of the induced E.M.F. in No. 3 is also on the point of changing.

Now in order to utilize the induced electromotive force to the best advantage the current must flow in every conductor in the same direction as the induced E.M.F., and therefore finally the commutator strip b must be under a brush.

For symmetry the commutator strip f must be under the other brush.

A moment's consideration will show that when the armature is running as a dynamo x is the positive brush.

The two paths in parallel from brush to brush are distinguished by thick and thin lines.

34. Formula for E.M.F. generated in a Dynamo.

To prove that in the armature of a dynamo the E.M.F. generated is given by

$$\frac{\mathsf{NP}n}{10^8} \times \frac{2}{b},$$

where N is the total magnetic flux threading the armature,

P is the total number of active conductors,

n is the number of revolutions of the armature per second,

b is the number of brushes.

Every conductor cuts the whole magnetic flux twice in every revolution.

If b is the number of brushes, there will be $\frac{\mathsf{P}}{b}$ active conductors in series.

Hence each 'series' of active conductors cuts $\mathsf{NP} \times \frac{2}{b}$ lines of force in every revolution, and the number of lines cut per second becomes $n\mathsf{NP} \times \frac{2}{b}$.

The average E.M.F. generated is therefore (by §§ 23, 26)

$$\frac{n\mathsf{NP}}{10^8} \times \frac{2}{b} \text{ volts.}$$

For a two-pole machine this becomes

$$\frac{n\mathsf{P}\,.\,\mathsf{N}}{10^8} \text{ volts.}$$

CHAPTER VII

CHARACTERISTICS OF MOTORS AND DYNAMOS

*35. Variation of Terminal D.P. of a Dynamo.

As the current of a dynamo is increased the D.P. between the terminals varies, but the variation is very different in the case of a 'series' dynamo from what it is in the case of a 'shunt' dynamo.

In any dynamo the total E.M.F. generated at any moment is given by $\frac{2\text{N} \cdot \text{P} \cdot n}{b10^8}$, where N is the total number of lines of force threading the armature, P is the number of conductors in the armature, n is the number of revolutions per second of the armature, and b is the number of brushes.

Hence in any armature if the magnetic flux is constant the total E.M.F. generated is proportional to the speed.

If the speed is constant the curve connecting the total E.M.F. generated with the current in the field circuit will be similar to the magnetisation curve for the iron of which the magnetic circuit is constructed (see p. 6).

Characteristic Curve of Series Dynamo.

On open circuit a small terminal voltage is obtained due to the residual magnetism of the field magnets.

As resistance is cut out in the external circuit and the current is increased, the total E.M.F. generated increases in proportion to the magnetic flux, and this of course varies with the current according to the magnetisation curve for the magnetic circuit.

While the iron is in a state corresponding to a point below the "knee" of the magnetisation curve, the total E.M.F. increases almost in direct proportion to the current, and a small change in the current usually produces a large change in E.M.F.

When the iron is approximately saturated there is practically no change in the total E.M.F. following upon a change in the current.

The terminal D.P. of a series dynamo is of course less than the total E.M.F. by a quantity equal to $C(R_a + R_m)$, so that after the total E.M.F. has reached its maximum value the terminal D.P. will begin to fall as the current is further increased, since the lost volts $C(R_a + R_m)$ are always proportional to the current.

To verify the above remarks a curve may be plotted connecting the terminal D.P. and current of a series dynamo. To compare the total E.M.F. with the current, the lost volts $C(R_a + R_m)$ must be added to the terminal D.P. in each case. A diagram is given which shows roughly the nature of the curves obtained.

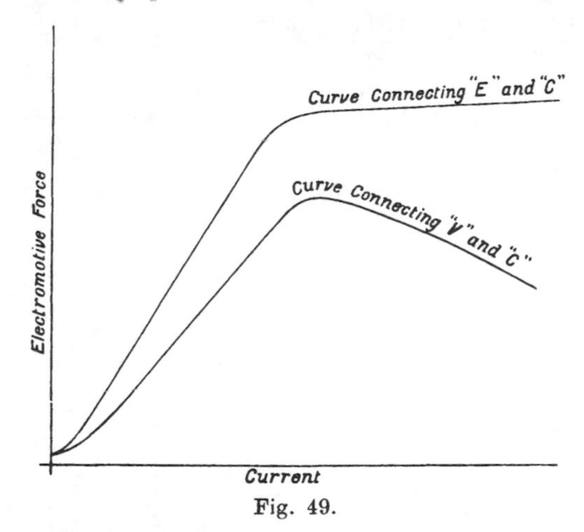

Fig. 49.

Characteristic Curve of Shunt Dynamo.

In considering the causes which bring about a change in the E.M.F. of a shunt dynamo when the current is varied, it is well to bear in mind the following facts:

In any particular machine the current in the field circuit depends solely on the terminal D.P.

Any change in the external current does not directly affect the current in the field circuit, it does so indirectly by affecting the terminal D.P.

There are no 'lost volts' in the field circuit.

The actions which take place are roughly as follows :

On open circuit the external current is zero and the armature current is the same as the current in the field circuit.

As resistance is cut out in the external circuit the external current increases and with it the armature current. This involves an increase in the lost volts, *e*, and a corresponding decrease in V and therefore in C_m which is proportional to V. The change in the magnetic flux consequent upon a fall in C_m will be large or small according as the state of the magnetic circuit corresponds to the 'steep' part of the magnetisation curve or to the more nearly horizontal part above the 'knee' of this curve.

The diagram represents a typical magnetisation curve, and

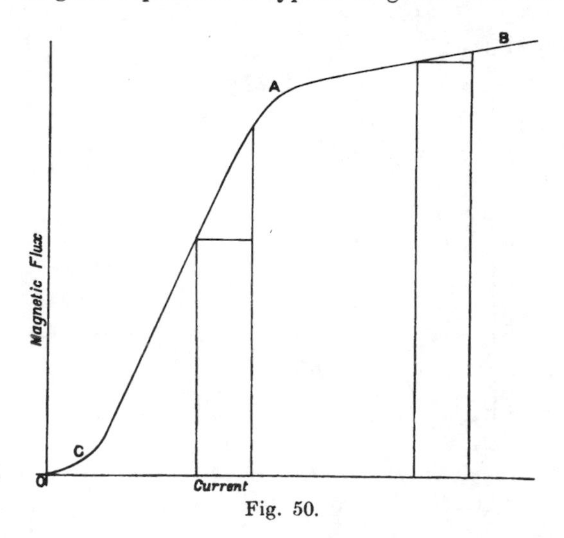

Fig. 50.

it is evident that a given change in the exciting current produces a much greater effect upon the magnetic flux when the iron is well below 'saturation' [i.e. between C and A] than when the iron is nearly saturated [i.e. between A and B].

A fall in the value of the magnetic flux involves a proportionate fall in the total E.M.F. E.

The ultimate fall in terminal voltage consequent upon an increase in the external current may therefore be diminished by employing highly saturated field magnets and by having an armature of very low resistance.

In a modern 'shunt' dynamo in which a high degree of saturation in the field magnets and a very low armature resistance are employed, the terminal D.P. remains very nearly constant for all loads from zero up to the normal working load.

A diagram is given which shows the nature of the two curves connecting

(i) Terminal voltage and external current,

(ii) Total E.M.F. and total current.

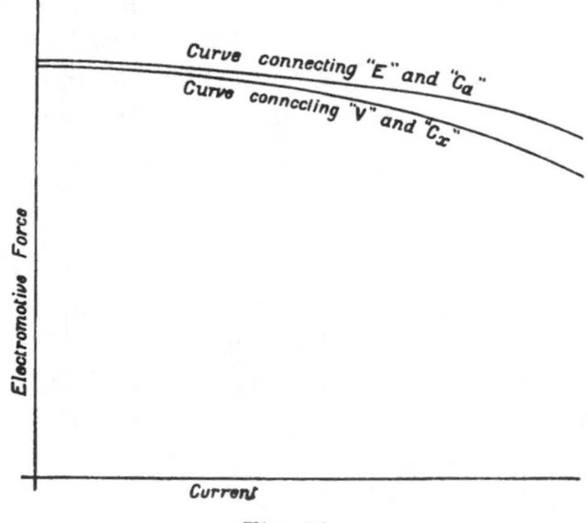

Fig. 51.

The two quantities V and C_x can usually be measured directly. In order to obtain E and C_a we have

$$C_a = C_x + \frac{V}{R_m},$$

$$E = V + C_a R_a.$$

It is sometimes desirable to use a dynamo whose terminal voltage can be readily varied, for example when the dynamo is to be used for battery charging.

When beginning to charge, resistance is put in to the field circuit, and as the charging proceeds this is gradually cut out.

In this way the flux and terminal D.P. are gradually increased and the current maintained at a constant strength in spite of the increasing counter E.M.F. of the battery.

For such a machine the field magnets are designed so that at normal working load their magnetic state corresponds to a point well below the 'knee' of the magnetisation curve, and thus a small variation in the resistance of the field circuit results in a considerable change in the terminal voltage of the dynamo.

When it is desired to transmit current to a distant station and to keep the D.P. at the terminals of that station constant, the dynamo which generates the current must be such that its terminal voltage rises with an increase in current in order to counterbalance the drop in potential along the mains. A 'shunt' dynamo is therefore used, round the field magnets of which are a few turns of conductor in series with the mains.

By adjusting the proportion of 'series' and 'shunt' turns round the field magnets the effect of the 'series' turns in increasing the magnetic flux as the current increases is made to over compensate the fall in the magnetic flux which would follow a rise in the external current if the machine were of the ordinary shunt type.

Such a dynamo is called compound or over-compound according as the voltage is absolutely constant with varying current or increases with increasing current.

*36. Behaviour of Motors under different conditions.

It will make the consideration of 'series' and 'shunt' types of motor simpler if we first consider the separately excited type.

Let E be the E.M.F. supplied to the brushes.

Let R_a be the armature resistance.

Let e be the back E.M.F.

Let C_a be the armature current.

The total power used by the armature is EC_a, and of this power is wasted in driving the current through the armature equal to $C_a^2R_a$.

The remainder is therefore converted into mechanical power which is given by

$$EC_a - C_a^2R_a.$$

Now $$C_aR_a = E - e,$$

hence the mechanical power $= eC_a$ and the efficiency therefore equals $\frac{e}{E}$.

Since $$C_a = \frac{E-e}{R_a},$$

another expression for the mechanical power is given by

$$\frac{e(E-e)}{R_a}.$$

We know that when a motor is running light the back E.M.F. is very nearly equal to the 'impressed' E.M.F. and that as the load on the motor increases, the back E.M.F. falls and the current increases.

We will now plot a curve showing how the mechanical power varies as e diminishes from the value E to zero (see Fig. 52).

[We will express e as a fraction of E and take E and R_a to be unity.]

From this graph it is clear that for values of e not less than 85 % of E, the mechanical power is very nearly directly proportional to $(E-e)$.

Generally speaking the back E.M.F. in the armature of a motor is never less than 90 % of the impressed E.M.F. at normal load and we may therefore assume without serious error that the mechanical power of a motor is directly proportional to $(\mathsf{E} - e)$ or in other words directly proportional to C_a.

e	$E - e$	Mechanical Power
1	0	0
·9	·1	·09
·8	·2	·16
·7	·3	·21
·6	·4	·24
·5	·5	·25
·4	·6	·24
·3	·7	·21
·2	·8	·16
·1	·9	·09
0	1	0

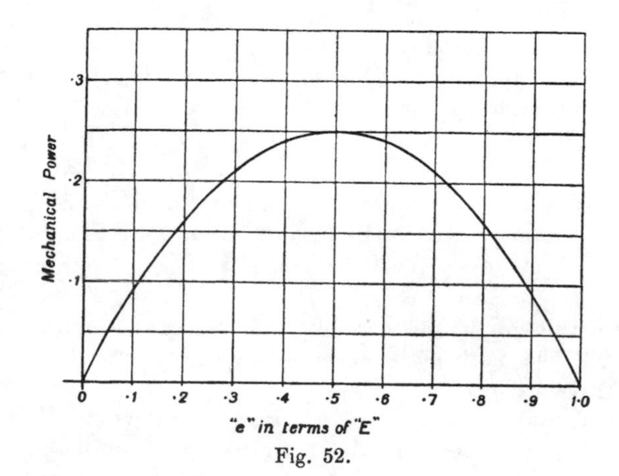

Fig. 52.

To obtain an expression for the torque exerted by the armature of a motor in terms of the armature current and flux we have

$$\text{Mechanical Power} = \frac{\text{Torque} \times 2 . \pi . n \times 60}{33000} \text{ H.P.},$$

where n is the number of revs. per sec., also Mechanical Power $= e . \mathsf{C}_a$, and $e = \frac{n\mathsf{PN}}{10^8}$ for a two pole armature, where N is the total flux and P the number of conductors in the armature; hence

$$\frac{n . \mathsf{P} . \mathsf{N} . \mathsf{C}_a}{10^8 \times 746} = \frac{\text{Torque} \times 2 . \pi . n . \times 60}{33000}$$

or $$\text{Torque} = \frac{\mathsf{C}_a . \mathsf{N} . \mathsf{P}}{10^8 \times 8{\cdot}5} \text{ lbs.-feet.}$$

When a motor is running, the torque, the magnetic flux and the impressed E.M.F. can each of them be varied independently of the others.

Consider the effect of varying any one where the others remain constant.

(i) Effect of varying the flux when torque and impressed E.M.F. are constant.

$$\text{Torque} = \frac{\mathsf{C}_a . \mathsf{N} . \mathsf{P}}{10^8 \times 8{\cdot}5} = \text{constant},$$

hence C_a varies as $\frac{1}{\mathsf{N}}$.

Also the mechanical power varies as C_a, hence mechanical power varies as $\frac{1}{\mathsf{N}}$. Since the torque is constant, the speed varies directly as the mechanical power, hence finally *the speed varies inversely as the magnetic flux.*

(ii) Effect of variation in 'torque' when the impressed E.M.F. and flux are constant.

$$\text{Torque} = \frac{\mathsf{C}_a . \mathsf{N} . \mathsf{P}}{10^8 \times 8{\cdot}5},$$

hence C_a varies directly as the torque. The mechanical power varies as C_a approximately, being really a little short of the value estimated on this assumption. Hence approximately the mechanical power varies directly as the torque.

Hence the speed is approximately constant. Really the speed falls very slightly.

(iii) Effect of variation in impressed E.M.F. when 'flux' and 'torque' are constant.

Since both 'torque' and 'flux' are constant C_a is constant.

$$C_a = \frac{E - e}{R_a}.$$

Hence $E - e$ is constant.

But e varies directly as n and $E - e$ is always very small compared with E.

Hence E varies directly as n or '*speed*' *varies as impressed* E.M.F. *when 'torque' and 'flux' are constant.*

Hence, to sum up, the speed varies as $\frac{E}{N}$, falling slightly with increasing torque.

*37. Shunt Motor.

A shunt motor may generally be considered as a separately excited machine whose field circuit is connected directly to a constant source of E.M.F.

We may therefore say at once that the speed will fall slightly as the torque increases.

The fact that the speed varies inversely as the magnetic flux is made use of in practice for regulating the speed of a shunt motor. A resistance is inserted in the field circuit which can readily be varied. The flux varies directly with the exciting current, though not as its first power, so that an increase in the resistance of the field circuit is followed by an increase in speed.

When considering the behaviour of a motor while starting, it is important to remember that the armature has not reached a steady state and that in fact the torque generated by the armature exceeds the torque due to the mechanical work which is being done externally by the belt or shaft, the difference or 'net' torque being applied to the acceleration of the armature. In practice the objects are:

(i) The generation of a large acceleration so that the motor may reach its normal speed quickly.

(ii) The prevention of unduly heavy currents which might injure the insulation of the conductors.

It is clear from this that on first connecting the motor to the supply of electricity the whole of the additional resistance in the field circuit must be cut out so that the current round the field magnets may be as large as possible.

Further a resistance must be inserted in the armature circuit to prevent a very heavy rush of current before the armature has gained speed and raised the back E.M.F. to its normal value.

It may be noted that in the case of a shunt motor, should the armature circuit be accidentally completed without additional resistance in series before the speed has got up (e.g. by putting the starting lever suddenly over) a very large current will flow, and moreover this large current will produce a considerable fall in the D.P. at the main terminals and consequently in the strength of the exciting current and of the magnetic flux.

The starting torque under these circumstances may scarcely exceed the normal torque in spite of the excess of current.

*38. Series Motor.

We will consider the two extreme cases:

(i) When the field magnets are saturated.

(ii) When the field magnets are in a state corresponding to a point below the knee of the magnetisation curve, i.e. when the flux is proportional to the current.

Effect of an increase of 'torque' on the speed.

(i) When the field magnets are saturated N may be taken to be constant, hence the speed varies as E', where E' is the D.P. between terminals of the armature circuit

$$E' = E - C_a R_m,$$

where E is the impressed E.M.F. (In a series machine $C_a = C_m$)

$$\text{Torque} = \frac{C_a \cdot N \cdot P}{10^8 \times 8{\cdot}5}$$

and, since N is constant,

C_a varies directly as the torque, hence the speed, which is proportional to E′, diminishes as the torque increases.

(ii) When the flux is proportional to the current

$C_a \cdot N$ varies directly as the torque.

C_a varies directly as N.

Hence N^2 varies directly as the torque or N varies directly as the square root of the torque.

We know that the speed varies as $\frac{E'}{N}$.

Hence speed varies as $\frac{E'}{\sqrt{\text{Torque}}}$.

In this case the effect on the speed is still more marked.

If a series motor is run light we have C_a very small and the torque very small.

E′ becomes equal to E and the speed becomes dangerously large.

In starting a series motor it is necessary to have a resistance in series with the armature on first connecting up, but even if the starting lever is accidentally put suddenly over the danger of burning out the armature is not so great as in the case of the shunt motor, since there cannot possibly be an excessive current in the armature unless at the same time the field magnets are highly saturated, and in this case the starting torque will be very large indeed and probably sufficient to raise the speed and back E.M.F. of the motor in time to prevent much harm being done.

CHAPTER VIII

THE PRODUCTION AND MEASUREMENT OF A MAGNETIC FIELD

39. The Magnetic Circuit.

An essential feature of a dynamo or motor is the magnetic field between the pole-pieces and the armature; the lines of force are produced by the current in the coils of the field magnets and conveyed to the armature by the iron of the magnets.

When we wished to determine the *electric current* in any conductor it was necessary to consider the whole circuit, taking into consideration the whole E. M. F., produced by cells, etc., and the various resistances in the circuit; in just the same way in order to determine the strength of the *magnetic flux* between pole-pieces and armature it is necessary to consider the whole 'circuit' through which the lines of magnetic force pass. It will be shown that there is a very close analogy between magnetic flux and electric current, the magnetic flux being produced by the field magnet coils and being less or greater according to the 'resistance' of the circuit traversed by the flux. In practical machines, such as a dynamo, the lines of magnetic force usually keep fairly close together, following a simple or divided circuit in the same way as does an electric current which flows through conductors of varying diameters, or through shunts, etc. It is reasonable then to talk of a *magnetic circuit.* But this is not always serviceable, as, for example, in the case of a bar magnet; in this the lines follow an indefinite number of paths through the air.

40. Magnetic Induction in Iron.

On page 2 it was pointed out that if a bar of iron is put into a solenoid in which a current is flowing, many more lines of magnetic force are created than were present before the iron was inserted; experiments such as those on pages 5 and 6 will show the great increase in 'pole strength' of a solenoid consequent on giving it an iron core.

The statement that a large number of lines of force emerge from one end of the bar, pass round through the air and enter at the other end is another way of saying that the iron becomes 'magnetised' by the coil; and in magnetised iron it is very probable that the molecules are arranged in lines instead of at random. Thus there is actual magnetism along definite lines through the iron, and we may suppose that the lines of force in the air are continued through the iron, so that they form closed curves; but inside the iron they are not called lines of force, but *lines of magnetisation* or *lines of induction*; and the latter term is often given also to the lines outside the iron. The lines of induction are commonly spoken of as the '*magnetic flux*,' and their number per sq. cm. is called the '*flux-density*'; this is always represented by the letter B.

Lines of induction may lie wholly inside iron; for example, we may magnetise a ring by winding a solenoid round it.

41. Measurement of Magnetic flux.

The simplest form of magnetic circuit consists of a uniform solenoid bent so that its ends meet, forming a circular ring; we will consider that the cross section of the solenoid is small compared with its diameter. There will be a magnetic flux round the ring which will be greater if the core is of iron than if it is of air; we will now consider a method of measuring this flux.

Assume that the ends of the solenoid do not exactly meet, but that there is a very narrow gap left. If there is an iron core, assume that this gap is filled by a moveable piece of iron of the same cross section as the iron core, scraped so as to make a good sliding fit. A thin coil of wire of several turns is to be made, at least as large in diameter as the coils of the solenoid, which can slip edgeways through the gap; in the case of the iron core, it will be wound on the moveable piece. The ends of this coil—called the 'secondary coil,' are to be connected to a galvanometer of special construction, called a 'ballistic' galvanometer. It is a mirror galvanometer, but its needle has a large moment of inertia, so that when a very brief current

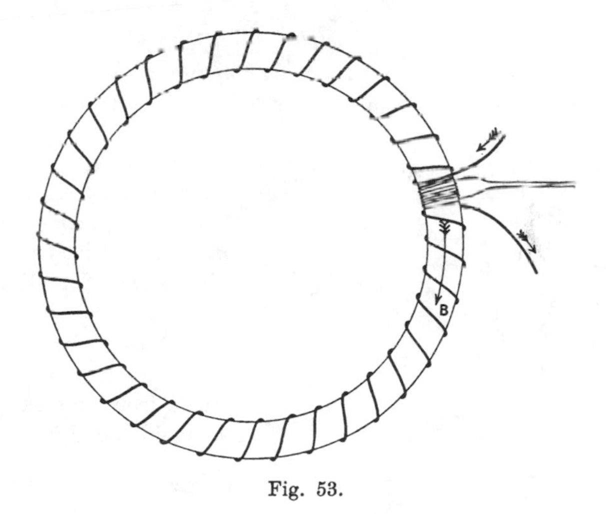

Fig. 53.

passes through the galvanometer, the needle has hardly time to begin to move before the current has ceased to flow. Under these conditions the *total quantity* of electricity that passes through the galvanometer is proportional to the angle through which the needle first swings.

Now suppose that N lines of magnetic force are by some means made to cut across each of the n turns of the secondary coil, in a very brief time, t seconds. They will induce an E.M.F. of $\frac{Nn}{t}$ absolute units, or $\frac{Nn}{10^8 \times t}$ volts, in the circuit during their passage. If we call the resistance of the galvanometer and secondary coil R ohms, the current produced will be

$$\frac{\mathrm{N}n}{10^8 \times t\mathrm{R}} \text{ amps.}$$

The quantity of electricity which flows in t secs. will therefore be

$$\frac{\mathrm{N}n}{10^8 \times t\mathrm{R}} \times t,$$

or

$$\frac{\mathrm{N}n}{10^8\mathrm{R}} \text{ coulombs.}$$

We see then that the angle of first swing of the galvanometer will be proportional to the number (N) of lines of force cutting

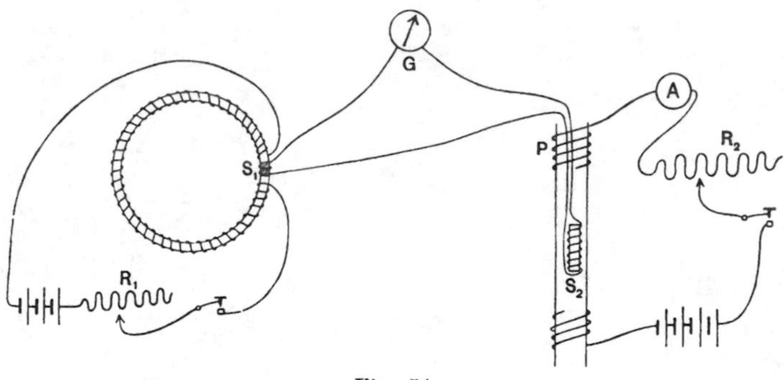

Fig. 54.

the secondary coil, since n and R are constant. So that we can determine a number which is proportional to the flux of magnetic force which traverses the secondary when it is in position in the solenoid, by snatching it away from that position, and observing the 'throw' of the galvanometer (since when in position all the flux in the solenoid traverses it, and when removed none of the flux does so).

In order actually to count the lines of force, we proceed to calibrate the galvanometer. In circuit with the secondary coil (S_1) and Ballistic Galvanometer (G) there is another secondary coil (S_2) of a large known number of turns wound on a short straight boxwood cylinder of known dimensions. This lies inside a long primary solenoid (P), through which we can pass a known current, measured by an ampère-meter (A). It will be shown in Art. 45 that we can calculate the number of lines of force per sq. cm. created within a straight solenoid when a current is started through it, and hence we can calculate the number of lines cutting across each turn of S_2 when a known current is created in P. A measured current is sent through P and the 'throw' of the galvanometer is observed; we can then deduce the number of lines of force corresponding to any throw. Both secondaries must remain permanently in the circuit, since the throw of the galvanometer depends *inter alia* on the resistance of the circuit.

We are thus provided with a fairly simple and accurate experimental method of measuring the total magnetic flux in this simple circuit.

42. Magnetomotive Force.

Consider a path along the axis of a ring solenoid, i.e. through the centre of each circular turn of wire; if the core is of iron, or any solid substance, imagine a very minute tunnel cut through it along the axis. Suppose that a unit magnet pole passes once round this path. We know that from a unit pole 4π lines of force emerge, so that 4π lines of force cut each turn of wire of the solenoid during one complete circuit of the pole. Suppose that there are n turns in the solenoid and a current of i absolute units is flowing, then by Art. 24, $4\pi ni$ ergs of work are done on the pole in its passage once round the circuit. The energy required is drawn from the electric current (or given to it, if the pole is forced round in the opposite direction). If we express the current in ampères (C amps say), then the work is $\frac{4\pi n\mathrm{C}}{10}$ ergs (since the same current is represented by i absolute units or $\frac{\mathrm{C}}{10}$ amps). This number may be represented by $\frac{4\pi}{10}\times$ ampère-turns.

Let us now turn to the analogy of the electric circuit. There is said to be unit Electromotive Force round an electric circuit when 1 erg of work is done by the electric forces during the passage round the circuit of a unit quantity of electricity (see p. 49). In the same way in the case of a magnetic circuit *there is said to be unit Magnetomotive Force round the circuit when 1 erg of work is done by the magnetic force of the current during the passage round the circuit of a unit magnet pole.*

Hence in the case we are considering, $\frac{4\pi n C}{10}$, or in general 1·257 times the number of ampère turns embraced, can be taken as the measure of the Magnetomotive Force (usually abbreviated to M.M.F.) round the circuit. This expression is quite general and holds for a magnetic circuit of any form; it has been explained for the case of a solenoid uniformly covering a ring, for the sake of simplicity, but a further discussion is given in Art. 51. It should be noted that the presence or absence of iron inside or outside of the solenoid makes no difference to the M.M.F., the only factor of importance being the number of ampère turns embraced by the path round which the M.M.F. is taken; there is no need for these ampère turns to be uniformly distributed along the path, the *total* M.M.F. being 1·257 times the *total* number of ampère turns embraced.

It is obvious that the magnetising power of the solenoid depends partly on the M.M.F. round it; in the case of electricity we made experiments to find out how the current in a given circuit depended on the E.M.F., and we can determine an analogous relation in the case of the magnetic circuit. This we will next proceed to do.

43. How the flux in a given magnetic circuit depends on M.M.F.

If a solenoid is wound uniformly round a ring of iron or steel, and if we increase the current step by step (and so the ampère turns and therefore the M.M.F.) we can measure the flux produced for each increase in M.M.F. The results for a ring made of mild cast steel are plotted in Fig. 55; the abscissae are not the actual M.M.F. for any particular length of solenoid, but the M.M.F. *per cm. of mean length* (usually called H), since it will be seen presently that the results can then be applied to a ring of any dimensions. Notice also that for the same reason the ordinates are not the total flux, but flux per sq. cm. (or B). The values of B for air have been added for comparison, but multiplied by 10 in each case for clearness.

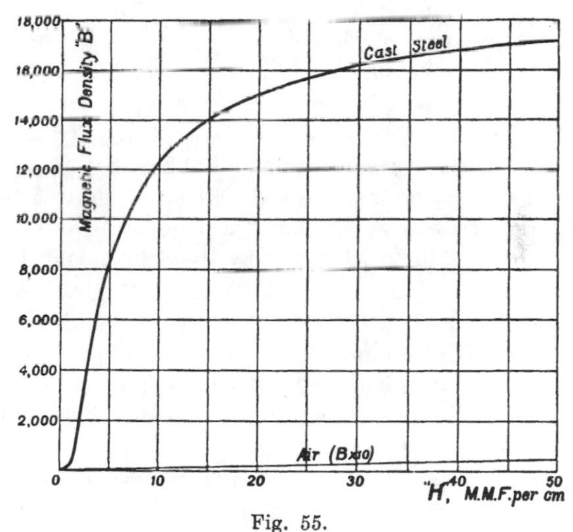

Fig. 55.
M.M.F. and flux for cast steel.

The curve for wrought iron is given on page 116.

It will be seen that the analogy with electric currents, etc., does not hold completely in the case of a magnetic circuit made of iron, since the flux is not proportional to the M.M.F. In the case of air, however, the analogy is complete.

It will also be seen that the flux in iron is enormously greater than that in air. To this property of iron a special name has been given, 'Permeability'; iron is said to be more permeable than is air to lines of magnetic force.

44. Permeability.

Definition of Permeability. If a solenoid is wound uniformily over a ring of uniform cross section, and of mean length l cm., made of any material, and if the total M.M.F. produced by the current in the solenoid when divided by l be termed H (the M.M.F. per cm.) and the consequent magnetic flux per sq. cm. be termed B; then *the value of the fraction* $\frac{B}{H}$ *is the permeability of the material.*

The permeability is always expressed by the Greek letter μ.

We will now prove that the flux density produced in a ring solenoid with an air core is equal to the M.M.F. per cm.

The flux density, B, is sensibly uniform all round the solenoid and at all points of its cross section; by the definition of lines of induction in air, the force on unit magnet pole is B dynes, and if the unit pole travels once round the ring, a distance of l cm., the work done is Bl ergs; therefore by the definition of magneto-motive force, the M.M.F. round the ring is Bl; therefore the M.M.F. per cm. is B, i.e. the M.M.F. per cm. is equal to the flux which it produces in air.

Thus for air, μ is unity, for all flux densities.

It is found that for all substances except iron (and steel, etc.), nickel and cobalt (which are chemically akin to iron) the permeability is very nearly unity; so that, magnetically, brass, zinc, wood, etc., are equivalent to air.

For any substance μ is a numerical coefficient and expresses the ratio of the permeability of the given substance to that of air, since for air $\mu = 1$.

It may now be seen why, in the definition of Permeability given above, the M.M.F. per cm. was termed H, the symbol usually reserved for the strength of a magnetic field. For if the medium with which we are dealing be air, we see that the M.M.F. per cm. is the actual magnetic force intensity; and if it be a magnetisable substance such as iron, we can show that the value of the effective magnetising force at a point in its interior is given by the M.M.F. per cm. *at that point.*

*45. Magnetic field in a straight solenoid.

We will next as an illustration consider the magnetic field inside a very long straight solenoid.

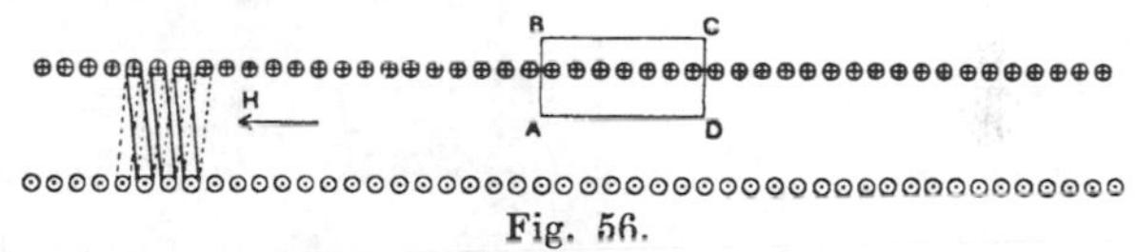

Fig. 56.

Suppose that Fig. 56 represents a section of the centre portion of a solenoid by a plane through its axis. Experiment and theory both show that practically the whole of the lines of force produced by a current in it leave the solenoid in the near neighbourhood of one end, and reach the other end by paths spread out through space in the same way as with a bar magnet. If then the solenoid is very long in comparison with its diameter, we can consider that it exerts no force at a point B or C just outside of it and near the middle of its length, as in the case of a long bar magnet; the "poles" at the ends from which the lines radiate are so distant as to have no sensible effect.

Again from symmetry we are justified in assuming that there is no force along a line AB or CD at right angles to the axis, whether the core be of air or of iron (or at least that if there were such a force it would be equal and in the same direction along both AB and DC).

Suppose the force intensity (i.e. the force on a unit pole) inside the solenoid parallel to the axis is called H. Then if we consider the forces acting on a unit pole while it moves round the rectangular path ABCDA, AD being l cm. in length, the total work done on or by the pole while it traverses this circuit is Hl ergs, since any work in AB and CD balances out, and no work is done in BC.

Now consider the M.M.F. round this path ABCDA. Suppose there are n turns of wire *per centimetre of length of the coil,* and that C ampères flow in it; then the number of ampère turns embraced in the path is nlC, so that the M.M.F. is $\frac{4\pi}{10} \times nl$C.

Equating these two values of the work done, we have

$$Hl = \frac{4\pi nlC}{10}$$

or

$$H = \frac{4\pi}{10} \times nC,$$

or in words, *the magnetic force intensity is* 1·257 *times the number of ampère turns per centimetre.*

This is true for the central portion of a solenoid whose length is practically infinite compared with its diameter, whether the core be of iron or air, etc.

46. Permeability of iron and steel.

It will be seen from Fig. 55 that the value of μ is not a constant for different values of the M.M.F. per cm. for any particular brand of iron; it also depends on the temperature and other conditions of the iron.

It is most convenient for practical purposes to know how the value of μ depends on that of the flux density for a given sample of iron, so results are often recorded in that form. They are derived from the curve in Fig. 55 by dividing the ordinate by the abscissa, and plotting the result as ordinate against the corresponding value of B.

A table obtained in this way is given opposite for various typical examples of iron and steel.

47. Saturation.

It will be observed that after a certain value, μ decreases with increase of B; if the magnetisation of the iron is pushed to a high value, it is found, as suggested in Fig. 55, that further additions to the M.M.F. do not increase the flux appreciably more than they would if the iron were changed to air. But it must be remembered that the *total* flux is still much greater than it would be if there were no iron present. In this condition it is as though the iron was 'saturated' with the lines of induction; a state which it is easy to explain on the molecular theory of magnetism by supposing that all the molecules have been faced round so as to lie along the lines of force (see Art. 55).

B	μ for grey cast iron	μ for annealed rolled plates for armatures	μ for cast steel	B
1000	244	1230	......	......
2000	289	1900	920	2000
3000	275	2320	1220	3000
4000	235	2600	1430	4000
5000	198	2750	1590	5000
6000	158	2800	1680	6000
7000	120	2760	1700	7000
8000	83	2590	1650	8000
9000	60	2300	1570	9000
10000	......	2170	1460	10000
11000	......	1930	1340	11000
12000	......	1710	1240	12000

These results for the cast steel of Art. 43 give the following curve:

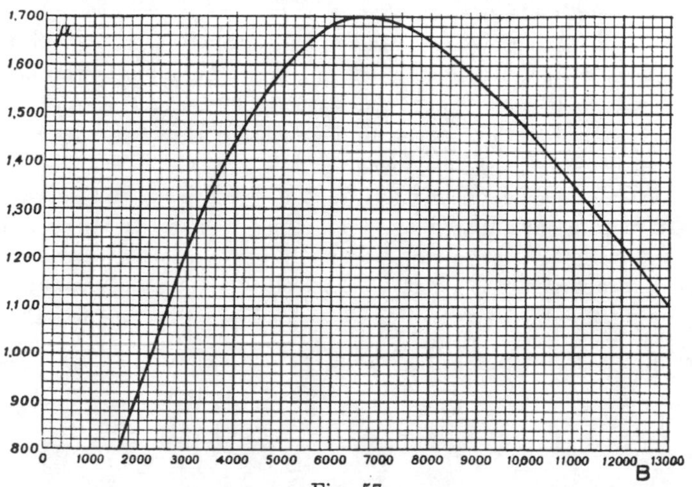

Fig. 57.

48. Reluctance of a Circuit.

In verifying Ohm's Law by experiment, we have to determine how the resistance must change with the E.M.F. in order to keep the current constant; we alter the resistance by altering the cross section or length of some of the wires in the circuit, or substituting manganin for copper, etc.

If we have a magnetic circuit, and vary the M.M.F. (by changing the current in the coils), we shall change the magnetic flux unless at the same time we suitably alter the cross section of the iron core to the solenoid, *or* change the length of the iron core along which the lines pass, *or* change the permeability by altering the material of which the core is made. On these three factors depends the flux which will be produced by a given M.M.F.; they determine a quantity analogous to the electrical resistance of a circuit; and this quantity is usually termed the magnetic *Reluctance* of the circuit. It can be defined directly as $\frac{\text{M.M.F.}}{\text{flux}}$ exactly in the same way as electrical resistance may be defined as $\frac{\text{E.M.F.}}{\text{current}}$.

We have now to discover how the reluctance depends on cross section, length and permeability.

In order to prevent confusion from the fact that the permeability of iron depends on the magnetic flux density through it, we must take as our object to determine how we must change the cross section, length or permeability so as to maintain the same flux *density* in spite of a change of M.M.F.

It is found by experiment that if we wish to preserve a constant flux density in an iron ring, in spite of alterations in M.M.F., we must alter either the *length* of the core in *direct* proportion to the M.M.F., or the *cross section* or *permeability* in *inverse* proportion to the M.M.F.—in other words, the reluctance is measured by the expression $\frac{l}{a\mu}$, where l is the length in cm., and a is the cross section in sq. cm.

So that the definition

$$\text{Reluctance} = \frac{\text{M.M.F.}}{\text{flux}},$$

leads to

$$\text{Flux} = \frac{\text{M.M.F.}}{\frac{l}{a\mu}}.$$

This is analogous to Ohm's Law, $C = \frac{E}{R}$, except that R does not depend on C, while μ depends on flux.

49. Practical applications of the Law.

To find ampère turns needed to produce a total flux of 30,000 lines round a ring made of cast steel of diameter 2 cm., whose section is circular and inside diameter is 20 cm.

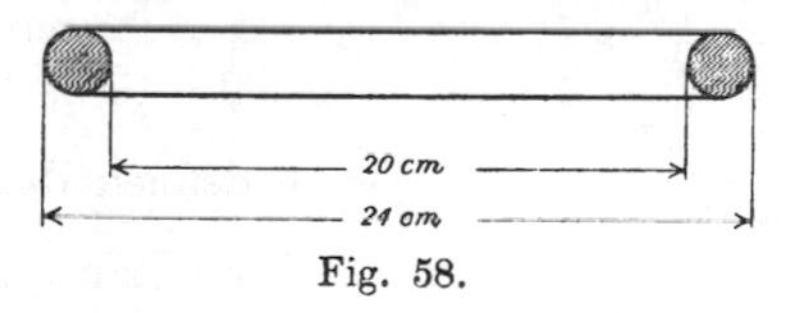

Fig. 58.

We have

$$l = \pi \times \text{mean diameter} = \pi \times 22 \text{ cm.}$$

$$a = \pi \times 1^2 = \pi \text{ sq. cm.}$$

$$\therefore \text{B} = \frac{\text{total flux}}{a} = \frac{30000}{\pi} \text{ lines per sq. cm.,}$$

and reluctance $= \frac{\pi \times 22}{\pi\mu}$.

We must now find what is the permeability of this brand of steel for a flux density of $\frac{30000}{\pi}$ lines per sq. cm.; suppose that it is steel quoted in the table, page 85; then we see by interpolation that the μ for this density is about 1520.

Thus we have $1{\cdot}257 \times$ ampère turns = M.M.F.

and M.M.F. = Reluctance × total flux.

$$\therefore \text{amp. turns required} = \frac{1}{1{\cdot}257} \times \frac{\pi \times 22}{\pi \times 1520} \times 30000,$$

or about 345.

We have hitherto considered a magnetic circuit consisting of a *uniform* ring of air or iron, but we can remove this restriction by extending the definition of Reluctance.

First suppose that the circuit consists of a series of rods of iron, of lengths l_1, l_2, l_3, etc. cm. and areas of cross section a_1, a_2, a_3, etc. sq. cm., and permeabilities μ_1, μ_2, μ_3, etc. Suppose the M.M.F. is provided by a magnetising coil round one or more of the bars, as in the figure.

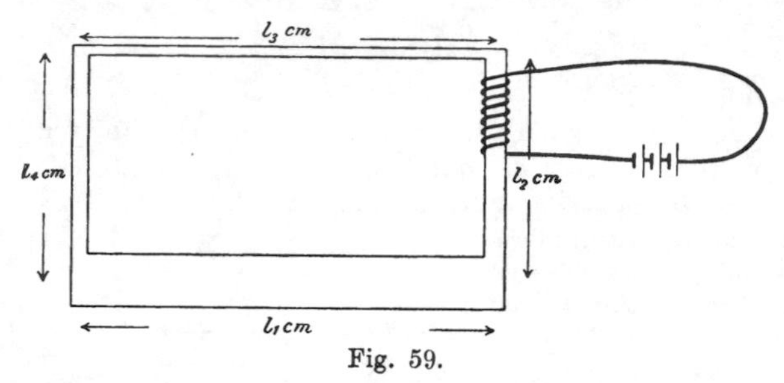

Fig. 59.

Then the reluctance of the whole circuit must be taken as

$$\frac{l_1}{a_1\mu_1} + \frac{l_2}{a_2\mu_2} + \frac{l_3}{a_3\mu_3} + \text{etc.}$$

Since the permeability of iron depends on the flux density, if we have a circuit consisting of rods made of steel of the same quality but with different cross sections, we must use this extended definition of reluctance in applying the law to such a circuit.

As an example of this, suppose we have a circuit composed of four cast steel rods of lengths 100, 40, 100 and 40 cm. and cross sections 1, 2, 4, 6 sq. cm. respectively, joined at the corners by blocks of iron whose reluctance we may neglect. Let us suppose that we wish to produce a flux of 12000 lines in the circuit. In the various bars the flux densities will be

$$\frac{12000}{1},\ \frac{12000}{2},\ \frac{12000}{4},\ \frac{12000}{6}\ \text{lines per sq. cm. respectively,}$$

and from the curve, as before, the respective values of the permeabilities will be 1240, 1680, 1220 and 920.

The total reluctance of the circuit is therefore

$$\frac{100}{1 \times 1240} + \frac{40}{2 \times 1680} + \frac{100}{4 \times 1220} + \frac{40}{6 \times 920},$$

or ·1203,

and the ampère turns needed to produce the flux of 12000 lines will be

$$\frac{1}{1 \cdot 257} \times 1203 \times 12000,$$

or <u>1148.</u>

We may note in passing that it is not quite so simple a matter to calculate the flux produced by a given number of ampère turns in such a circuit, since each of the various values of μ depends on the total flux produced. It is necessary to proceed by assuming a likely value of the flux and calculating as above the ampère turns needed to produce it, and repeating the process until we have a curve from which to deduce the flux corresponding to the given ampère turns. Contrast this with the simple process of calculating the current produced in an electric circuit by a given E.M.F.

50. Effect of an air gap in the circuit.

Suppose that in the above circuit, the bar whose cross section is 6 sq. cm. is sawn through so that there is a gap of 1 cm. in the middle of it. We may consider that we still have a magnetic circuit, but that the flux has to traverse a disc of air, 6 sq. cm. in cross section and 1 cm. thick, the permeability of which is unity. As a matter of fact the flux will not pass quite straight across the gap, but will bulge out somewhat, but with so short a gap this will not seriously affect the result.

With this assumption, the reluctance of the whole circuit will be

$$\frac{100}{1 \times 1240} + \frac{40}{2 \times 1680} + \frac{100}{4 \times 1220} + \frac{39}{6 \times 920} + \frac{1}{6 \times 1},$$

or ·2534;

and the ampère turns required will be

<u>2419.</u>

So that the introduction of the narrow air gap has more than doubled the ampère turns needed to produce the same flux.

*51. Leakage of Flux.

We have assumed in the last article that the presence of an air gap in the circuit does not affect the M.M.F. round that circuit. A thoughtful student may not be prepared to accept this; he may object on the grounds that the ends of the iron on each side of the gap are strongly magnetised, and that from these ends spring out lines of force which must have an influence on the unit magnet pole in its passage across the gap, and perhaps through the iron, in traversing the circuit, an influence which did not exist when the iron formed an endless ring. To show that this does not affect the *total* M.M.F. round the circuit, we will now discuss the value of the M.M.F. round a circuit of the most general kind.

Consider any field of magnetic force, containing permanent magnets and pieces of soft iron or other materials; first, let us assume that no electric current is flowing anywhere in the field.

Consider any path through this field, forming a closed curve or circuit; if it passes through iron, etc., a narrow tunnel must be imagined for the path. Although it is not necessary for the argument it will be easier to think of this path as following an existing line of force. Now imagine a unit magnetic pole to pass slowly once round the circuit. Although sometimes it will be impelled onwards and sometimes retarded by the magnetic forces, yet in performing the whole circuit there can be no balance of work either done on the pole or done by the pole; for if there were, by allowing the pole to move round and round the circuit in the direction in which it is found to give a balance of useful work, we should have an inexhaustible supply of energy, which is contrary to experience as summed up in the Law of Conservation of Energy.

Next, consider such a field containing also a circuit in which a current of electricity (i absolute units) is flowing. Suppose the path of the pole is linked once with this current; then (by Art. 24), since 4π lines of magnetic force proceed from the pole, these 4π lines of force will cut the current bearing conductor, and work to the amount of $4\pi i$ ergs will be done by or on the pole as it passes once round its circuit, the energy being drawn from the current. If its path is linked n times with the current, $4\pi ni$ ergs of work will be done. If, however, the path of the pole is not linked with any current, no balance of work will be done in spite of the presence of the current in the field.

If W ergs of work are done on the unit magnet pole in its passage once round the circuit, W is the measure of the Magnetomotive Force round that circuit. Therefore *the M.M.F. round a circuit is measured by* $\frac{4\pi}{10}$ *times the number of ampère turns embraced by the circuit.* This expression can be used for any circuit whatever.

We were therefore correct in disregarding the induced magnetism of the iron when calculating the total M.M.F. round the circuit of Art. 50.

The assumption as to the value of the reluctance was, however, not so completely justifiable; the expression for it can only be used when the whole flux is confined to a definite and complete "circuit." This was approximately true in that case, but would not be at all true if, for example, the whole of one bar had been removed; and it is not possible to amplify the definition of Reluctance so that such a case may be included under the law which we have given connecting Flux, M.M.F. and Reluctance.

If we map out with iron filings or an exploring compass the field round any magnetised bodies (except an endless uniform ring wound uniformly with its magnetising solenoid) we shall find that more or less lines of force leak out from the sides of the iron, and that those which leave an end do not proceed as though they were enclosed in a cylindrical tube, which is the assumption we have made above.

For example, in the case of a straight solenoid with an iron core, the flux spreads from the neighbourhood of one end throughout the whole of space in its passage towards the other end of the solenoid, even backwards through the iron core itself, thus to some extent demagnetising the iron core.

But if this solenoid is bent into a ring so that its ends nearly touch one another, the great majority of the flux passes practically straight across the narrow air gap, and we can then employ the idea of the magnetic circuit if we allow for the air gap in the expression for reluctance*.

The law we have given for a simple circuit will therefore not hold in the general case any more than Ohm's law would for an electric circuit consisting of wires immersed in a badly conducting liquid such as a solution of copper sulphate, with breaks in the metallic circuit where the current had to pass through the liquid.

In a practical circuit such as that of a dynamo or motor this objection holds to a greater or less extent; but the gaps are small, consisting only of sufficient space for the armature conductors and their clearance. These gaps help to cause a leakage of the flux, so that the whole of that generated by the ampère turns in the coils of the field magnets does not reach the armature (which is where it is required). This leakage in practice is not calculated, but determined experimentally for any required design of the machine, and the requisite increase in ampère turns provided for.

52. Method of Measuring Permeability.

The result on page 89 shows that the joint in the iron to some extent vitiates the method of measuring permeabilities described on page 76; we will now describe a method free from this objection.

Take a continuous ring of the material to be tested; wind the magnetising coil uniformly all round it; at some point wind on the ring under the magnetising coil a secondary coil of say 20 turns in series with a ballistic galvanometer and another secondary within a solenoid as before. Now increase the current in the magnetising solenoid step by step, by cutting out resistances in this circuit, and note the throw of the galvanometer at each increase. These throws measure the increased number of lines of induction created in the ring by the increased M.M.F., since

* For a further discussion, see Appendix on p. 108.

each line has to cut across all the turns of the secondary; for they end by linking with the secondary and previously they did not exist. We can then plot a curve like that of fig. 55, giving the flux density corresponding to the M.M.F. per cm., or the Magnetisation Curve for the material under test*.

53. Hysteresis in Iron.

Take a ring of wrought iron that has not been magnetised, and take its magnetisation curve while increasing the M.M.F. as

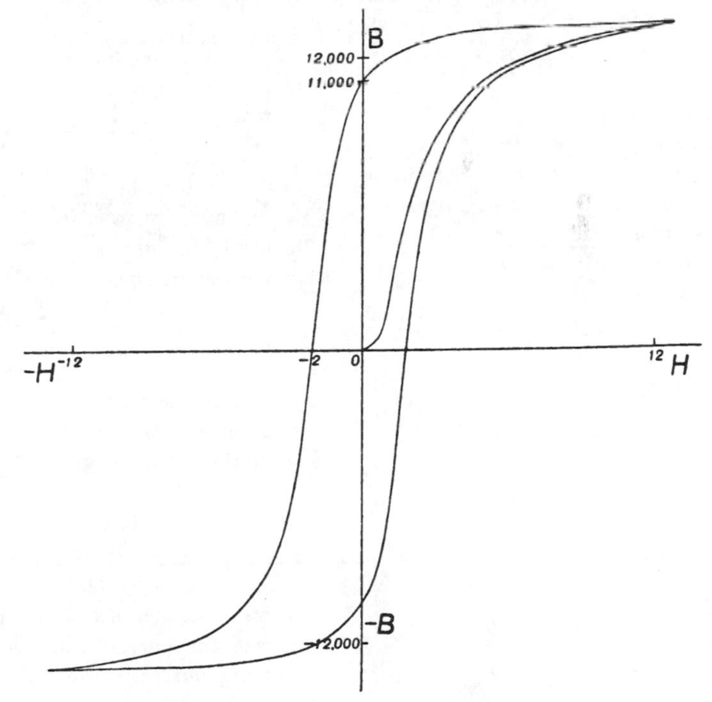

Fig. 60. Hysteresis Curve for Wrought Iron.

described above, pushing the magnetisation so far that the iron approaches saturation. Next insert resistances in the

* For numerical results, see Examples 34 to 41.

circuit of the magnetising current, so reducing the M.M.F. step by step, and note the decrease of flux density at each step. We shall find that the curve is not the same as before; the magnetisation does not decrease as rapidly as it rose, for a given change of M.M.F., and when all M.M.F. is removed there is some residual magnetisation. Now reverse the magnetising current, and increase this reversed current step by step as before; then decrease it until it is zero, reverse again and increase it, step by step, so performing a complete cycle. On plotting the results, we shall obtain a curve as shown in Fig. 60.

From a study of this curve we see that the magnetisation 'lags behind' the magnetising force, and to this property of iron Professor Ewing gave the name 'Hysteresis' (from the Greek ὑστερέω, 'I lag behind').

Different qualities of iron show this property to a greater or less extent; mild steel only slightly more than wrought iron, cast iron more still, while hard steel possesses it to a high degree, as will be seen from this diagram.

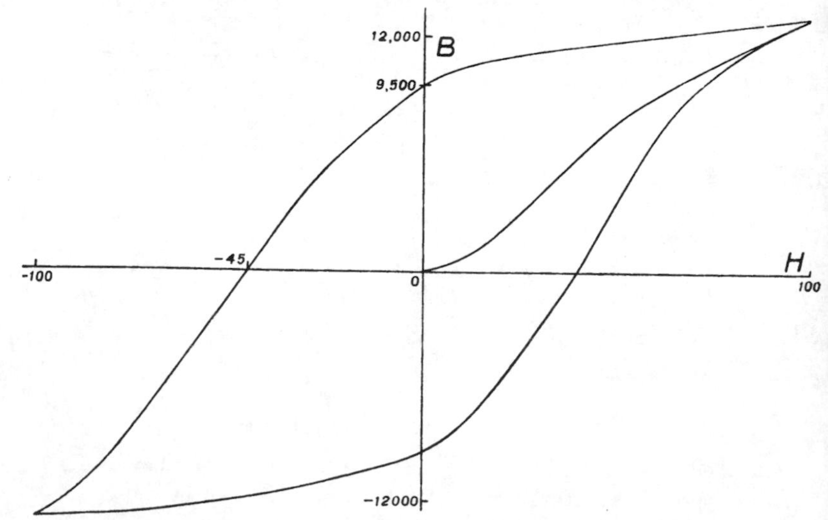

Fig. 61. Hysteresis curve for hard steel wire.

It will be seen that hard steel possesses considerable residual magnetism, hence it is used for permanent magnets. In order to destroy the residual magnetism, we have to apply a reverse magnetising force of 45—to this force the name 'Coercive Force' is given, since it represents the force with which the material holds its residual magnetism.

54. Energy Loss in Magnetisation.

It can be shown that the area of the loop in the magnetisation curve represents the energy required to carry the material through the cycle of magnetisation; this energy all reappears as heat in the material magnetised and is drawn from the magnetising current. For such purposes as the core of an armature, which passes through a complete cycle of magnetisation at every revolution, it is clearly desirable to select iron which gives a small amount of hysteresis, as both the loss of energy and the heating are objectionable; for the pole pieces of a dynamo or motor whose armature is slotted it is also important to use wrought iron, but for the field magnets this is not necessary; for them only a high permeability is required, since the magnetisation in them does not go through any cycle.

It will be seen that the permeability of iron does not depend only on the value of the M.M.F., or even on the flux density at the moment in the iron, but also on the previous 'history' of the iron. It is even possible to have a *negative* value to μ, as between the points marked 11,000 and -2, in figure 60.

55. Molecular Theory of Magnetisation.

We will now give Professor Ewing's hypothesis as to the physical reason for the behaviour of iron, etc., under magnetising forces.

We may regard a piece of unmagnetised iron as built up of a multitude of molecules, each a permanent magnet, grouped at random; and we may look on the process of magnetisation as a 'combing out' of this tangle; the more strongly the iron is magnetised, the greater is the number of these tiny magnets that fall into line. Complete 'saturation' of the iron occurs when all the molecules are pointing along the lines of induction.

These minute magnets are supposed to be free to turn and take up a new direction, as is a pivoted compass needle; but they are situated so close to one another that they are constrained more or less to remain in one position by the influence on them of their neighbours. This can be illustrated by laying down about a hundred small 'exploring compasses' side by side on a table with their cases in contact; the needles will be seen to settle down in definite positions, about which they oscillate at first, and the general effect is that there is no particular direction in which a majority seems to point. If now a coil of wire be so arranged that a uniform magnetising force can be applied to the compass needles by passing a current through the coil, and if we first pass a very small current, the effect will be to deflect slightly some of the compass needles towards the direction of the impressed lines of magnetic induction.

If we now decrease the magnetising current to zero again, the needles all return to their initial positions.

But if we gradually increase the magnetic field by increasing the current, a new stage occurs at which one of the needles breaks away from the group which holds it, and of which it forms part. This may cause two or three neighbouring needles in the group to break away from one another, and their change of position may slightly affect the needles in neighbouring groups, but the effect will not on the whole be considerable. On decreasing the current to zero again, the initial condition will not be regained, as the broken group will not be reformed.

On increasing the current further, the breakdown of groups will go on very rapidly, until all the needles are dominated by the magnetising force, though the influence of their neighbours still keeps them from lying exactly along the lines of induction; but further considerable increases of the magnetising force makes this more nearly the case.

We can easily see how clearly, on this hypothesis as to the

molecular condition of iron, these results account for the behaviour of iron under magnetising forces.

Observations made on iron in very weak magnetic fields show that the flux density, B, is very nearly proportional to the magnetising force, H. In other words, the curve in Fig. 55 connecting B and H, starts by being a straight line while very near to 0. It has been found* for hard iron that for values of H less than 1·2, the following formula holds for μ, $\mu = 81 + 64\,H$; so that μ starts by being 81 and may be considered constant for values of H less than ·01. The beginning of the magnetisation curve for hard iron may therefore be represented thus, if the scale be greatly magnified.

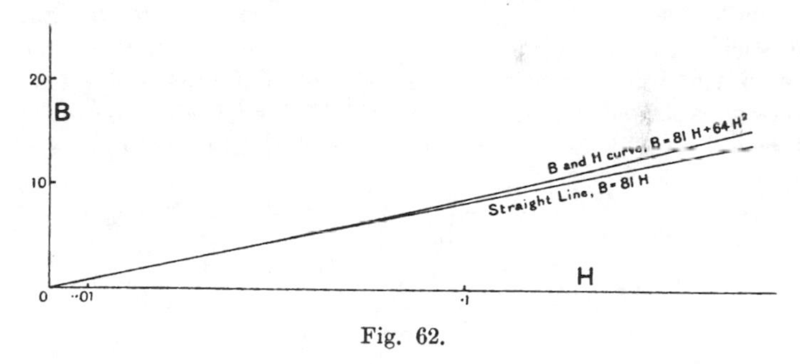

Fig. 62.

For such small magnetising forces it is found also that there is no residual magnetism, and no hysteresis when the magnetisation is carried through a complete cycle. This stage corresponds to the first stage in our experiment with compass needles, when they were slightly deflected but no group was broken up, and on removal of the magnetising force, the original condition was restored. It is exactly analogous to the behaviour of materials under stress when within the elastic limit.

If now we increase H, we get to the steep part of the curve of magnetisation, when the groups of molecules rapidly break down —and if we remove the magnetising force these groups do not

* See Ewing's Magnetic Induction in Iron, p. 121.

reform, and we obtain residual magnetism and hysteresis. This is analogous to the behaviour of material stressed beyond the elastic limit; and the loss of energy in carrying the magnetism of a piece of iron through a complete cycle is analogous to the loss in straining a piece of material beyond its elastic limit, while the energy expended in straining it is all restored if the strain does not go beyond that limit.

We also see how the 'knee' and the comparatively horizontal part of the magnetisation curve are accounted for by the hypothesis, as representing the behaviour of the material after all the groups are broken down. One point it is important to note, that even when all the molecules are exactly aligned with the lines of induction, yet B will increase with H; for B represents the number of lines per sq. cm., and will increase with H even if there is no iron present to be magnetised, or if that iron is completely saturated. So that we must not expect the curve ever to become actually horizontal, but to run parallel to the 'air' curve.

CHAPTER IX

REACTIONS IN THE ARMATURE

56. Armature Reaction.

We have seen that in the case of both dynamo and motor, when any active coil or conductor is moving parallel to the direction of the lines of force in the field, that is when it is reversing the direction in which it is cutting lines of force, then the commutator strip to which it is connected must be in contact with a brush.

The line through the centre of the armature perpendicular to the direction of the magnetic flux is called the *neutral axis*, the reason for the name being that a conductor which is on this axis is generating no E.M.F.

The line through the centre of the armature which cuts those conductors which are directly connected to the brushes is called the *axis of commutation*.

We shall see presently that the effect of self-induction in the conductors necessitates the axis of commutation being displaced slightly from the neutral axis, but for the present we shall neglect the effect of self-induction.

When no current is passing through the armature the position of the neutral axis is the same as that of the axis of symmetry between the two poles, but when a current passes through the armature it generates a field of force in a direction (if the axis of commutation coincides with the axis of symmetry) at right angles to the direction of the field set up by the field magnets alone.

The diagrams refer to a ring armature. Fig. 63 shows the nature of the field generated by the field magnets alone, when there is no current in the armature; AA is both axis of symmetry and neutral axis. Fig. 64 shows the nature of the field generated

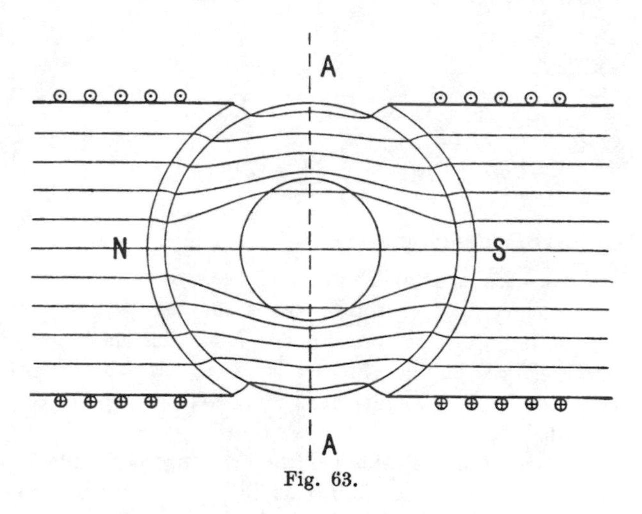

Fig. 63.

by the armature current alone. AA is the axis of symmetry, BB the neutral axis. The direction of magnetisation of the armature core in Fig. 63 is at right angles to that in Fig. 64.

When a dynamo or motor is loaded the armature core is subjected to the action of the two fields of force described above, at the same time.

The resultant of these two fields of force is a field of force inclined to the direction of the axis of symmetry.

For example suppose AB in Fig. 65 represents in direction and magnitude the field due to the field magnets and BC that due to the armature current.

[It may be remarked here that the field of force due to the armature current is always small by comparison with that due to the field magnets.]

Then AC represents the resulting field of force both in direction and magnitude. A diagram (Fig. 66) is given showing the combined effect of the two fields in the case of a dynamo.

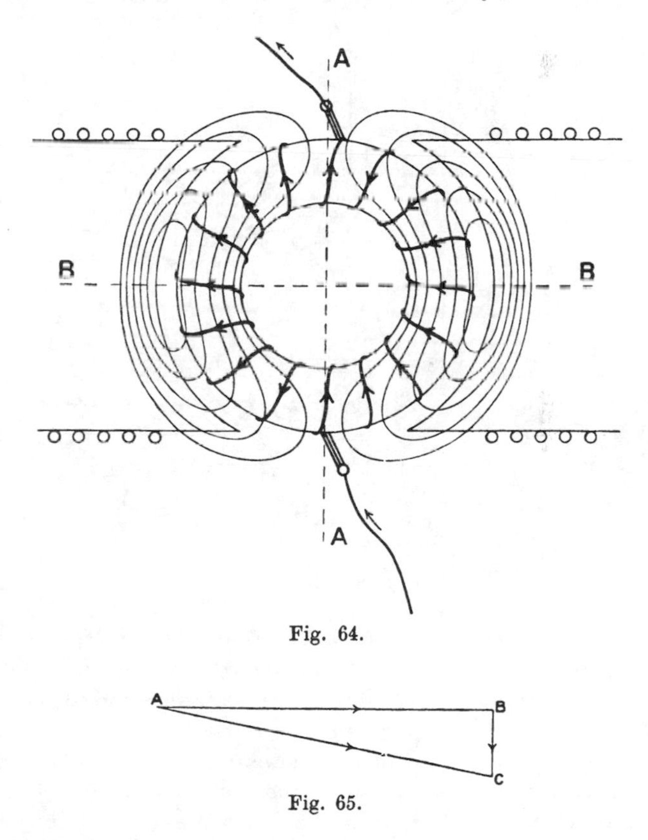

Fig. 64.

Fig. 65.

The neutral axis is displaced from the axis of symmetry in the direction of rotation. The axis of commutation, and therefore the brushes, must consequently be moved round in the direction of rotation.

AA is the axis of symmetry, BB the axis of commutation and neutral axis.

The corresponding diagram for the case of a motor should be drawn, and it will be found that for the same direction of

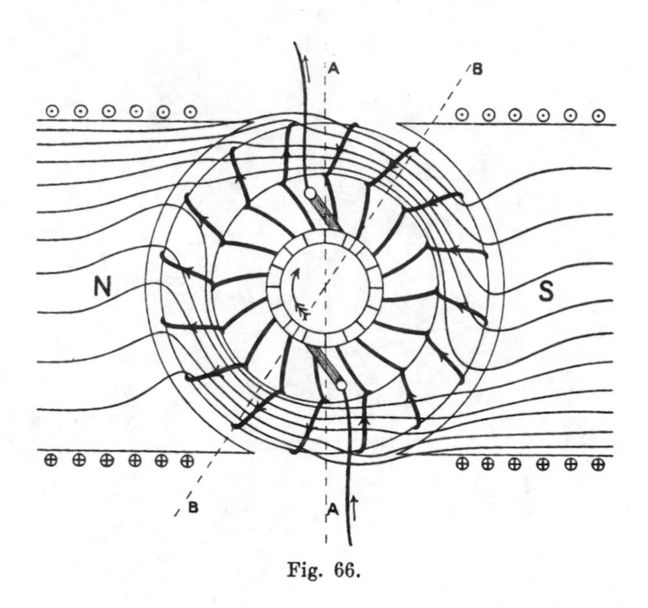

Fig. 66.

rotation the armature current is reversed, the direction of the cross magnetising field is therefore reversed and hence the neutral axis is displaced against the direction of rotation instead of with it as in the case of the dynamo.

When the brushes are moved forward or back so as to bring the axis of commutation on to the neutral axis, the direction of the magnetising field due to the armature current becomes inclined to the direction of the axis of symmetry, and reference to Fig. 67 shows that this produces a still greater distortion of the combined field, that the brushes have in fact to be moved until the axis of commutation comes up with the neutral axis and this will occur when the direction of the field of force due to the armature current is perpendicular to the resultant field of force.

ABC is the triangle of forces corresponding to the case when the axis of commutation lies on the axis of symmetry.

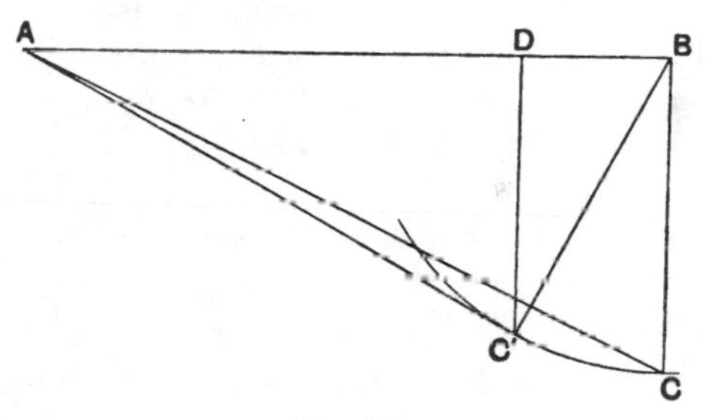

Fig. 67.

ABC′ is the triangle for the case when the axis of commutation is brought on to the neutral axis.

The neutral axis advances through the angle CAC′, while the brushes advance through the angle CBC′.

In the figure ABC′ draw C′D perpendicular to AB, then BC′, which represents the field of force due to the armature current, may be resolved into components BD and DC′. The component BD tends simply to diminish the field due to the field magnets, while the component DC′ tends to alter the direction of the field.

BD is called the demagnetising component and DC′ the cross magnetising component of the field due to the armature current.

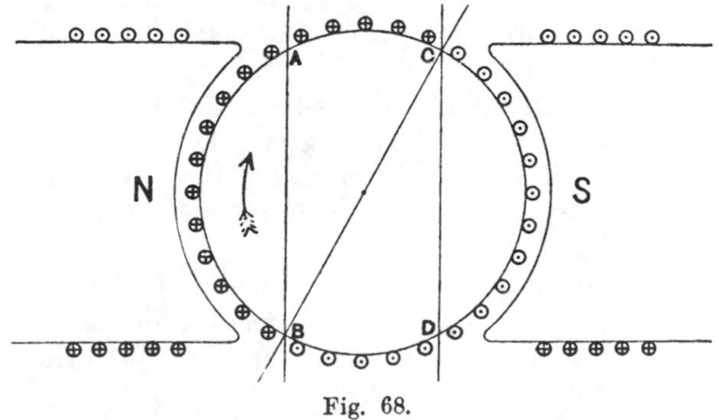

Fig. 68.

It should be noticed that in the case of a dynamo the flux density in the forward horn of each pole piece is greater than that in the hindward horn and vice versa in the case of a motor.

The winding of an armature may be divided up into two parts, one part consisting of coils which produce a cross magnetising field only and the other of coils which produce a demagnetising field only.

In Fig. 68 BC is the axis of commutation, BA and CD are drawn parallel to the axis of symmetry.

All conductors included between A and C and between B and D tend simply to demagnetise the field magnets, while the remainder tend simply to cross magnetise or alter the direction of the field.

57. Self-Induction.

When a single conductor carries a current and the circuit is suddenly broken, it is found that an electromotive force is set up in the conductor which tends to keep the current flowing in the same direction, and if a conductor is suddenly connected to a source of electromotive force a back electromotive force is generated in the conductor which prevents the current rising immediately to its full value.

The E.M.F. referred to is called the E.M.F. of self-induction and may be explained as follows:

Consider the conductor as being divided in half along its length, i.e. as consisting of two parallel conductors. Then the stopping of the current in one half of the conductor generates a momentary E.M.F. in the same direction in the other half.

In fact we may think of the magnetic field linked with the current as falling in on the conductor when the current is stopped and in doing so generating an E.M.F. in the same direction as the original current.

Again the starting of a current in one half of a conductor generates a momentary E.M.F. in the opposite direction in the other half of the conductor, with the result that the rise of the current to its full value is retarded.

This effect will clearly be increased if the conductor is wound into the form of a coil, as a large proportion of the lines of force generated by any one turn will cut all the other turns as well as itself in the act of being created or destroyed.

Further if such a coil has an iron core the strength of the field which is linked with the current will be very much greater than if there were no core and the circuit will be still more highly self-inductive.

An instructive experiment on this subject may be performed with the aid of a strong electromagnet of the horseshoe type.

Close the magnetic circuit of the electromagnet by placing an iron yoke across the poles of the magnet, then if the current is stopped after the electromagnet has been excited the magnetic lines will remain in the iron until the magnetic circuit is broken by knocking away the iron yoke. Connect a voltmeter across the terminals of the exciting coils of the magnet and knock away the iron yoke. The magnetic field collapses, and in doing so cuts all the turns of the exciting coils and induces an E.M.F. in them which can easily be read on the voltmeter.

58. Sparking.

We have seen that, neglecting the effect of self-induction, the axis of commutation of an electric machine must coincide with the neutral axis, and that the neutral axis is displaced from the axis of symmetry in the direction of motion in the case of a dynamo and against the direction of motion in the case of a motor.

The diagram (Fig. 69) refers to a dynamo, OS being the axis of symmetry, ON the neutral axis coinciding with the axis of commutation.

Now we know that if a circuit is carrying a current and the E.M.F. which produces the current is suddenly removed the current continues to flow owing to the fact that the collapse of the magnetic field which was allied with the current sets up an E.M.F. which tends to keep the current flowing in the same direction as before, and we know on the other hand that if an electromotive force is suddenly introduced into a circuit the current in that circuit cannot rise to its full value at once because the generation of the magnetic field in connection with the current sets up a counter E.M.F. which opposes the flow of the current. The name of self-induction was given to these phenomena.

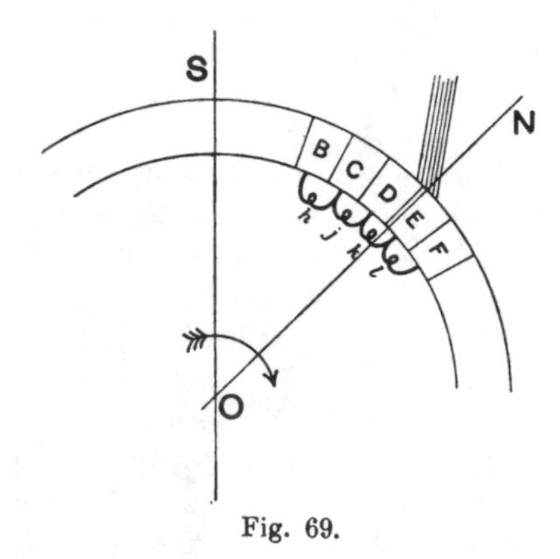

Fig. 69.

Referring to the diagram we see that the coils h, j and l are each carrying half the total current of the armature. The coil k is short circuited, and since D has only just come under the brush the coil k is at the moment generating an E.M.F. of self-induction in the direction of DkE. The coil k is not generating any E.M.F. due to cutting the magnetic flux since it is on the neutral axis; and therefore a current is flowing from D through k to E and back through the brush.

Now as soon as E leaves the brush the coil k will have to carry half the total armature current in the reverse direction, and the generation of this current in k also produces an E.M.F. of self-induction in the direction DkE.

With such an arrangement in practice this E.M.F. of self-induction is considerable, and just as the strip E leaves the brush there is a virtual momentary break in the circuit (since the coil EkD is not prepared to form part of this new circuit); the current in the part of the armature forward of the brush then takes the only course open to it by forming an arc from E to the brush. This appears as a spark and is destructive of the commutator.

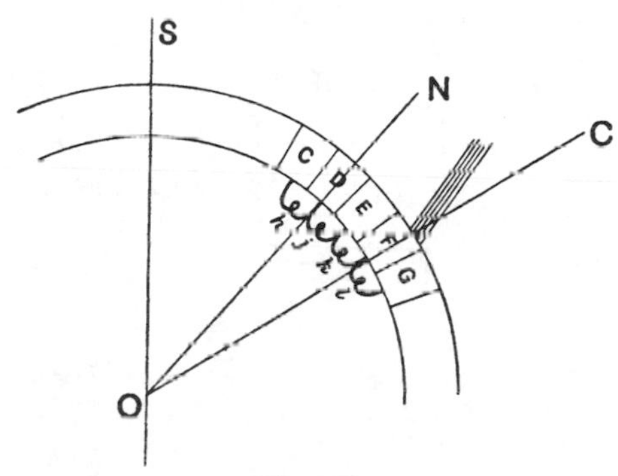

Fig. 70.

To overcome this sparking the axis of commutation is advanced still further to the position OC (Fig. 70).

Now during the time that the coil l is short circuited it is generating an E.M.F. in the direction GlF since it has passed the neutral axis and is cutting the magnetic flux in the opposite direction. If the axis of commutation is advanced through the correct angle this E.M.F. destroys the self-induced current in l and creates a reverse current equal to half the armature current by the time l is thrown on to the right-hand side of the armature. If this occurs there will be no self-induction in l as G leaves the brush and consequently no sparking.

The correct position of the brushes is found by trial and it should be noticed that unless there is a strong field in the region of the coil l while it is short circuited a sufficient reverse E.M.F. will not be produced to prevent sparking.

Now the reaction of the armature on the field magnets tends to weaken the field near the hindward pole horn and under certain circumstances it becomes necessary to introduce compensating coils to prevent this reaction, otherwise the sparking may be excessive.

APPENDIX

NOTE ON THE USE OF MAGNETIC CIRCUITS

Consider a closed electric circuit containing cells A, A′ in series producing the current, one set B being charged, a resistance R, an arc lamp L, and a motor M. Let *a*, *b*, *c*, ..., *z* be points on the circuit a yard apart.

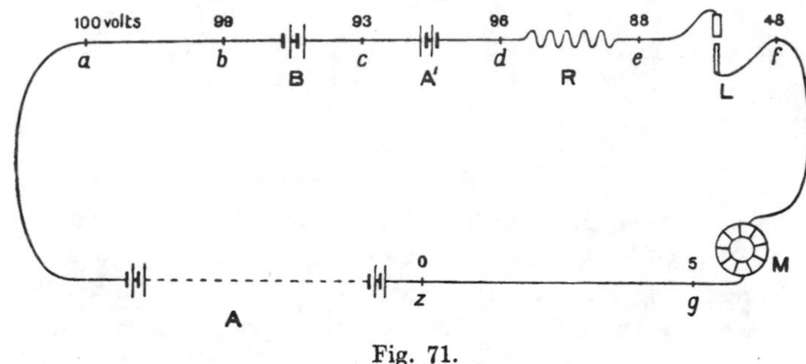

Fig. 71.

Suppose that a voltmeter connected to *z* and to *a*, *b*, *c*, etc. in succession shows voltages as marked in Fig. 71; these are probable values assuming that the E.M.F. needed to drive the current in the circuit through a yard of the conducting cable is 1 volt.

We observe that the "E.M.F. per yard" differs widely at different parts of the circuit, and is nowhere equal to its average value; but we can find the current by applying Ohm's Law either to the whole circuit or to any of the portions *ab*, *bc*, etc.

Next consider a magnetic circuit, consisting of a cast steel ring with an air gap, surrounded by a complete ring solenoid. Suppose that the length of the ring is 30 cms., its cross sectional area 2 sq. cms., the width of the gap 1 mm., and that the flux density is 7000, the permeability corresponding to this being 1700 (see Fig. 57).

The reluctance of the whole circuit is $\frac{30}{2 \times 1700} + \frac{\cdot 1}{2 \times 1}$ or ·059, so that the M.M.F. needed to produce a flux of 2×7000 lines will be $2 \times 7000 \times \cdot 059$ or 824. Therefore the average M.M.F. per cm. is $\frac{824}{30}$ or **27·5.**

The flux density in the air gap, and therefore the M.M.F. per cm. there, is **7000.**

The flux density in the iron is 7000, and $\mu = 1700$, therefore the M.M.F. per cm. in the iron is $\frac{7000}{1700}$ or **4·12.**

We see then that in a magnetic circuit the M.M.F. per cm. may change from point to point, as may the E.M.F. per cm. in an electric circuit, and that it may differ throughout from its average value.

Since the "M.M.F. per cm. in the iron" is defined to be the work done by the current in the coil in forcing a unit pole a distance of 1 cm. along a tunnel, filled with air, cut through the iron, and since this tunnel passes completely round the ring thus forming a complete circuit in air, it may

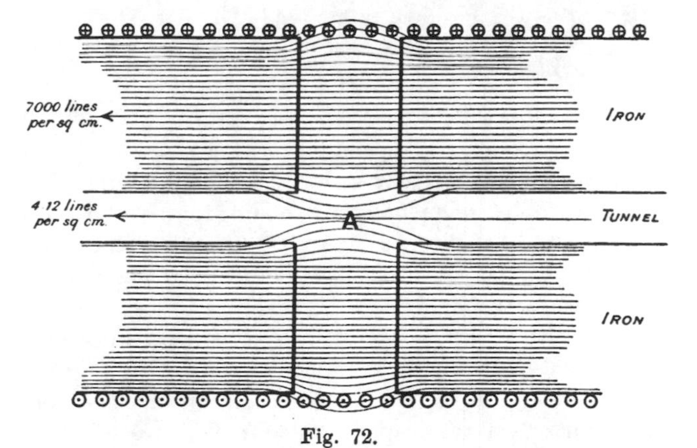

Fig. 72.

not be obvious why there should be sudden changes between 7000 and 4·12 in the M.M.F. per cm. in this air, instead of 27·5 throughout. This may be clear if we consider the field of force in the neighbourhood of the air gap, as sketched in Fig. 72, where the tunnel (greatly magnified) is shown entering the iron.

As the tunnel is indefinitely reduced in diameter, the flux density at A opposite the end of the tunnel will approach 7000 lines per sq. cm. as a limit, and it will be seen that, in consequence of the great permeability of the iron, very few of these 7000 lines per sq. cm. at A succeed in entering the tunnel, but that they crowd into the iron by preference. The flux density at A is therefore much greater than in the tunnel; and as the flux density is, in air, numerically equal to the M.M.F. per cm., therefore inside the iron the M.M.F. per cm. is much less than in the air gap.

If the air gap had not existed, the total M.M.F. round the iron would still have been 824, and the M.M.F. per cm. would have been $\frac{824}{30}$, or 27·5, throughout; the presence of the air gap alters the distribution of this M.M.F., so that 700 exists in the narrow air gap, and 124 in the iron. The air gap has increased the reluctance, and a large part of the M.M.F. is employed in "driving the flux" through the air gap; but the air gap does not alter the total M.M.F. round the circuit, any more than the introduction of a resistance in an electric circuit alters the total E.M.F.

In a case where we can treat the flux as following such a simple circuit as this, we can safely apply the law

$$\textit{Total}\ \text{flux} = \frac{\textit{Total}\ \text{M.M.F. round circuit}}{\text{Reluctance of circuit}},$$

just as we use Ohm's Law for a whole circuit.

But the expression corresponding to Ohm's Law for part of a circuit, however truly we may state it, is of no practical service since we have no instrument corresponding to a voltmeter with which to measure "difference of M.M.F." (except in the case of a uniform field in air, when a vibration magnetometer could be used to measure H).

In electricity we are not practically concerned with currents which can leak out of the conductors through an unlimited volume of badly conducting medium (except in the case of "earth returns" in telegraphy, which experience shows have negligible resistance; we cannot assume zero *reluctance* for an "air return," and it is not true); we are also possessed of insulators.

In magnetism, on the other hand, most of the practical problems involve the former difficulty, as for example in short electromagnets; and we have no medium which will not transmit lines of induction, with which to surround our iron. Hence all solutions by this method must be approximate, and we must employ such devices as leakage coefficients even when dealing with favourable problems (except in such extreme cases as we have discussed above, in which the whole circuit might have been considered to be surrounded with an "insulator," since we assumed no leakage to take place).

There is, however, another method of attacking the problem of magnetic induction, which was formerly the only one used and which is applicable in many cases where the use of the idea of a magnetic circuit is impracticable, e.g. a short straight magnet. This depends on the consideration of the magnetic force at individual points, as a resultant of the "external" force and the actual state of magnetisation of the iron, together with the "intensity of magnetisation" produced by that force at those points. On the basis of these two quantities can be built up many of the properties of magnetic induction which we have explained, as will be found in text-books dealing with the general phenomena of magnetism; this process lends itself to the mathematical treatment of general problems. The advantage of the "magnetic circuit" is that it provides us with a simple method of solving many practical problems, with an accuracy sufficient for practical purposes, which would be impossible by the other method. For example, we have seen that we can get a rough value for the flux through the armature of a dynamo produced by a current flowing in the field circuit; it is rough because we can assign only a rough value to the leakage, and because the assumption of equality in the length of all lines of induction in the circuit is certainly not true; as a consequence, the M.M.F. per cm. at different points of the cross section of the iron is different, and therefore the flux-densities at those points. But it would be a hopeless task to try to calculate H at all points of the circuit even roughly, in order to deduce the intensity of magnetisation; hence the method of the magnetic circuit is alone developed in this book, which does not touch on the questions where the other method is preferable.

EXAMPLES

In these examples, $\frac{4\pi}{10}$ *may be taken as* $\frac{5}{4}$, *and* $\frac{10}{4\pi}$ *as* ·8.

1. A series motor has an armature resistance of ·3ω and a field circuit resistance of ·8ω.

Calculate the electrical efficiency of the machine when supplied with 100 volts. and taking a current of 10 ampères. 89 °/₀.

2. What is the commercial efficiency when delivering 1·1 mechanical H.P.? 82·06 °/₀.

3. A shunt dynamo has armature resistance ·2ω and field circuit resistance 50ω.

Calculate the electrical efficiency when delivering an external current of 50 amps. at a terminal D.P. of 100 volts. 87·12 °/₀.

4. Calculate the electrical efficiency of the same machine when used as a motor supplied with a terminal pressure of 100 volts. and taking 50 amps. current. 86·78 °/₀.

5. A shunt dynamo has an armature resistance of ·25ω and a field circuit resistance of 200ω. The dynamo generates 240 volts. on open circuit at 1000 revs. per min. Assuming that the field magnets are saturated when the exciting current is above 1 amp., calculate the terminal D.P. when giving an external current of 20 amps. 234·7.

6. What is the electrical efficiency in this case? 92·37 °/₀.

7. How many lines of force emerge from a magnet pole of 20 units strength? 251·36.

8. A magnet is 20 cms. long from pole to pole and the pole strength is 10 units. What is the strength of the magnetic field at a point outside the magnet and on the axis distant 5 cms. from one pole

(*a*) Neglecting the effect of the more distant pole

(*b*) Including the effect of the more distant pole?

(*a*) ·4 lines per sq. cm. (*b*) ·384 lines per sq. cm.

9. A magnet AB is one metre long, D is a point between A and B 20 cms. from A, DC is at right angles to AB and is 20 cms. long. Find the direction of the magnetic force at C.

39° 34′ to AB.

10. A conductor 2 metres long carries a current of 20 amps. at right angles to a uniform straight magnetic field of 200 lines per sq. cm. What force in grammes weight acts on the conductor?

204.

11. A current of 5 amps. flows along a conductor of 3ω resistance. How many ergs of work are done in 2 minutes?

9×10^{10}.

12. A conductor 2 metres long lies horizontally and carries a current of 200 amps. The "earth's vertical force" being ·42, what will be the force in grammes weight tending to move the conductor horizontally? 1·7 grm. wt.

13. A conductor on a drum armature is 20 cms. long and carries 30 amps. The magnetic field in the air gap is 3000 lines per sq. cm.; find in lbs. wt. the force on the conductor.

·4 lb. wt.

14. A conductor 5 cms. long moves at 20 cms. per sec. across a field of force of 1000 lines per sq. cm.

What E.M.F. is generated in the conductor? ·001 volt.

15. A conductor 2 metres long moves at 100 cms. per sec. across a field of force of 2000 lines per sq. cm. The ends of the conductor are joined by a wire whose resistance is 3ω, the resistance of the conductor itself being negligible.

Find the E.M.F. generated and the force which opposes the motion of the conductor. ·4 volt. 5333 dynes.

16. A conductor 30 cms. long whose resistance is $\cdot 2\omega$ is free to move across a magnetic field of 1500 lines per sq. cm. and is supplied with an E.M.F. of 30 volts.

Find the initial force on the conductor and the final speed attained. 675,000 dynes, 66,666 cms. per sec.

17. A two-pole drum armature consists of 500 active conductors and is threaded by a flux of 10^6 lines.

The armature is rotated at 1000 revs. per min.

Find the E.M.F. generated. 83·33 volts.

18. The armature described in Question 17 has a resistance of ·1ω. It is used as a motor supplied with a terminal P.D. of 100 volts.

Find the speed attained

(*a*) when running light,

(*b*) when absorbing 5 kilowatts.

(*a*) 1200 revs. per min. (*b*) 1140 revs. per min.

19. The armature of a four-pole four-brush motor consists of 400 active conductors and is threaded by a flux of 10^7 lines.

The resistance of the armature is ·05ω.

Find the speed and torque of the motor when supplied with an E.M.F. of 240 volts. and a current of 40 amps.

94·1 lb. ft. 714 revs. per min.

20. A shunt motor is connected to 240 volt mains. Resistance of field circuit 200ω. Resistance of armature ·2ω.

The total current is 10 amps.

Calculate the armature current, back E.M.F. and electrical efficiency. 8·8 amps. 238·24 volts, 87·36 °/ₒ.

21. A circular cast steel ring, rectangular in section, is wound evenly with 100 turns of wire. The inside diameter is 15 cms., the outside 17 cms.; the depth is 2 cms. Determine, using Fig. 57, the current needed to produce a flux density of (*a*) 3000, (*b*) 4000, (*c*) 5000, (*d*) 13,000 lines per sq. cm.

(*a*) ·984, (*b*) 1·11, (*c*) 1·26, (*d*) 4·73 amps.

22. Hence calculate the approximate total flux produced by a current of 1·15 amps. 8700 lines.

23. A hard steel wire of diameter 2 mm. and length 60 cms. is put into a very long straight solenoid which has 80 turns per cm. and a current of 1 amp. flowing in it. From the curve of Fig. 61, page 94, we find that the corresponding flux density is 13,000 lines per sq. cm. Calculate the total number of lines of induction in the wire. 408.

24. Remembering that from unit pole there emerge 4π lines of induction, and assuming that all the lines emerge from points very close to the end of the wire, calculate the strength of the pole at each end of the wire. 32·5.

25. If the current is now stopped, calculate from Fig. 61 the number of lines of induction remaining in the wire, and hence the strength of the permanent poles. 298; 23·75.

26. What current must be sent in the reverse direction to demagnetise the wire completely? ·45 amps.

27. Find μ for a flux density of 6000 lines per sq. cm. for wrought iron, whose magnetic qualities are given in Fig. 73.

$\mu = 2{,}500$.

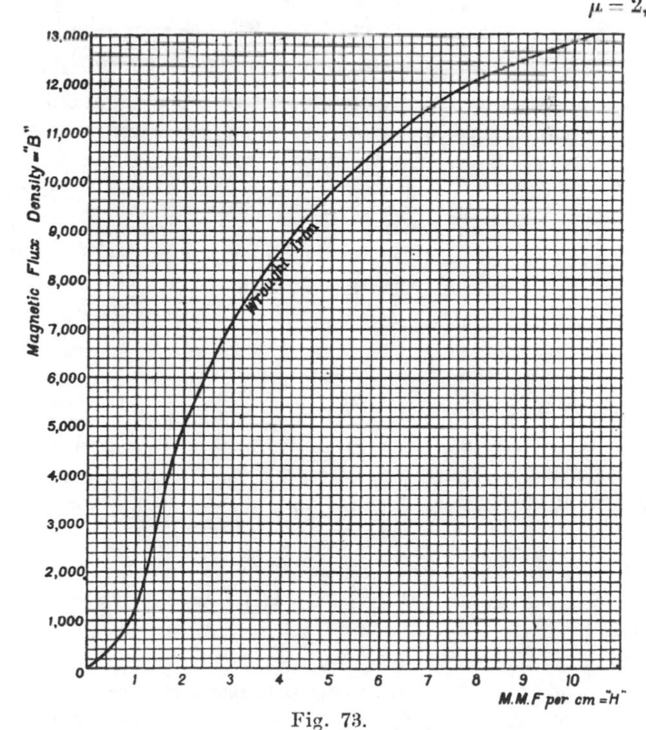

Fig. 73.

28. A ring solenoid in the form of a circular coil of mean length 30 cms. and cross section 1·5 cms. contains 120 ampère turns; calculate the M.M.F. per cm. round it, and deduce the total flux within it if the core be of air. 5 ; 7·5.

29. Suppose that the core is formed of the wrought iron of Question 27, find the total flux. 14,700.

30. Suppose that the core is formed partly of air and partly of a ring of the wrought iron, the length of the ring being 30 cms. and its cross section ·2 sq. cm. Calculate (*a*) the flux density and (*b*) the total flux in the iron and (*c*) the total flux in the air remaining within the solenoid, remembering that in each substance

$$H = \text{M.M.F. per cm.} = \frac{4\pi}{10} \times \text{ampère turns per cm.}$$

(*a*) 9,800 ; (*b*) 1,960 ; (*c*) 6·5.

31. A long solenoid of circular section has a mean diameter of 1 cm. and 20 turns of wire per cm. of its length. Find H within it when it carries 5 amps.; also find the total flux through it. 125, 98·2.

32. On inserting an iron wire of 1 mm. diameter, the total flux through the coil is 258 lines; find the flux density in the iron and the permeability for this flux density.

$B = 20{,}500$; $\mu = 164$.

33. What current will be needed to produce a magnetic field of intensity 200 at the centre of a coil (with air core) 100 cms. long, wound with 500 turns of wire. 32 amps.

34. A coil of 935 turns of copper wire is wound on a pasteboard tube of circular section and 144·7 cms. long; calculate the number of lines per sq. cm. at its centre when 1 amp. flows in the wire. H = 8·12.

35. A coil of 963 turns of copper wire is wound on a cylinder of boxwood 20 cms. long of diameter 4·94 cms., the diameter over the wire being 4·99 cms. A uniform field of H lines per sq. cm. parallel to the axis is suddenly created within the coil; calculate the total number of lines linked with the copper wire.

18,653 × H.

36. If the coil in Question No. 35 is placed centrally in the coil of No. 34, and a current of 1 amp. is started in the latter, calculate the number of lines of force linked with the former coil.
$1{\cdot}505 \times 10^5$.

37. If the inner or secondary coil in No. 36 is coupled to a ballistic galvanometer, and the throw of the galvo. on *reversing* a current of 1 amp. in the outer primary is observed to be 169 scale divisions, calculate the number of lines of induction which must be linked with, or unlinked from, the circuit of the galvo. to produce a throw of 1 scale division. 1781 lines.

38. If a circular cast-iron ring, mean circumference 34·7 cms., has 231 turns of wire wound on it, calculate the value of the M.M.F. per cm. (or H) for 1 amp. flowing in the wire.
H = 8·4 lines per sq. cm.

39. If a secondary coil of 14 turns is wound on this ring, in circuit with the ballistic galvo. of Question 37, and a throw of 80 scale divisions is observed on starting a current of 5 amps. in the primary coil (of 231 turns), calculate the total flux produced in the ring. 10,180 lines.

40. If the sectional area of the iron of the ring be 1·7 sq. cm., calculate B in Question 39. 5,988 lines per sq. cm.

41. From the results of Questions 38 and 40, calculate the approx. value of the permeability of the cast iron for a flux density of 6000 lines per sq. cm. $\mu = 142$ approx.

42. If a gap 1 mm. wide is sawn through the ring of Question 40, calculate the reluctance of the circuit for a flux density of 6000 lines per sq. cm. ·202.

43. What current must now be sent through the 231 turns to produce a flux density of 6000 lines per sq. cm.? 7·08 amps.

44. If the ring had been made of the cast steel whose magnetic qualities are given in Fig. 57, what current would be needed through the 231 turns to give a flux density of 6000 lines per sq. cm. (*a*) in the solid ring, (*b*) when the gap of 1 mm. had been cut in it? (First find the corresponding μ.)
(*a*) ·43 amp. (*b*) 2·51 amps.

45. Calculate the reluctance of a soft iron ring of mean diameter 40 cms. and cross section 4 sq. cms., the value of μ being 1900. ·0165.

46. Calculate the reluctance of the same ring cut into two semicircular parts which are held with the ends 1 mm. apart. ·0665.

47. Find in each of the above two cases the M.M.F. necessary to produce a flux of 40,000 lines. 660, 2660.

48. Hence find the number of ampère turns needed on each ring. 528, 2128.

*49. In a certain motor the armature consists of 252 conductor bars and there are two brushes.

The following observations connect the total effective flux (N) with the field circuit current C_m :—

C_m	N
·5	$8{\cdot}6 \times 10^5$
1·0	$15{\cdot}3 \times 10^5$
1·5	$20{\cdot}25 \times 10^5$
2·0	$23{\cdot}5 \times 10^5$
2·5	$25{\cdot}4 \times 10^5$
3	$26{\cdot}5 \times 10^5$

The resistance of the field circuit is 46ω and that of the armature $\cdot185\omega$.

Calculate the speed of the motor when running light at a terminal D.P. of 120 volts, 80 volts and 40 volts, also when absorbing 3 kilowatts at 80 volts.

Ans. 1. 1112 revs. per min.
,, 2. 864 ,, ,,
,, 3. 700 ,, ,,
,, 4. 789·4 ,, ,,

INDEX

The numbers refer to the pages

The numbers refer to the pages

Printed in the United States
By Bookmasters